Reliable Software for Unreliable Hardware

Semeen Rehman • Muhammad Shafique
Jörg Henkel

Reliable Software
for Unreliable Hardware

A Cross Layer Perspective

 Springer

Semeen Rehman
Karlsruhe Institute of Technology
CES – Chair for Embedded Systems
Karlsruhe, Germany

Muhammad Shafique
Karlsruhe Institute of Technology
Department of Computer Science
Karlsruhe, Germany

Jörg Henkel
Karlsruhe Institute of Technology
Department of Computer Science
Karlsruhe, Germany

ISBN 978-3-319-25770-9   ISBN 978-3-319-25772-3   (eBook)
DOI 10.1007/978-3-319-25772-3

Library of Congress Control Number: 2015960246

Printed on acid-free paper

This Springer imprint is published by Springer Nature
The registered company is Springer International Publishing AG Switzerland

Preface

Embedded systems have become an important part of our day-to-day life because of their pervasive deployment in various application domains such as consumer electronic devices like smartphones and tablets, medical healthcare, telecommunication, automotive, aircrafts and space-applications, etc. Due to the shrinking transistor dimensions, embedded computing hardware is getting increasingly susceptible to different reliability threats like *transient faults* (such as soft errors due to high-energy particle strikes) and *permanent faults* due to design-time process variations and run-time aging effects. Therefore, reliability has emerged as one of the primary design criteria in the nano-era. Soft errors manifest as spurious bit flips in the underlying hardware that may jeopardize the correct software execution. However, design-time manufacturing process variability manifests as frequency and leakage power variations in different cores in a multi-/manycore processor, while run-time aging effects result in frequency degradation over the period of time. A majority of state-of-the-art hardware-level reliability techniques employ full-scale redundancy or error correction blocks in processor components that result in significant overhead in terms of area, performance, and/or power/ energy, which may be prohibitive within the stringent design constraints of embedded systems. To alleviate the overhead of hardware-level techniques or targeting the low-cost unreliable hardware, several software-level reliability improving techniques have evolved that are based on the concept of full-scale redundancy at the code or data level and, therefore, incur significant performance and memory overhead ($\geq 2\times$–$3\times$). In general, state-of-the-art software-level reliability techniques have, by far, not exploited their potential since the common belief, so far, was that reliability problems when occurring at the hardware level should be addressed at the hardware level. However, a lot of hardware-level faults can potentially be masked at the higher software layers. Furthermore, a manycore processor is subjected to multiple reliability threats that need to be considered when providing functional and timing correctness.

To enable a highly reliable software system for embedded computing, this manuscript introduces novel concepts, strategies, and implementations to leverage multiple system layers in an integrated fashion for reliability optimization under user-provided tolerable performance overhead constraints. To enable this, this work addresses the key challenge of *bridging the gap between the hardware and software* by quantifying the effects of hardware-level faults at the software level while accounting for the knowledge of the processor architecture and layout. It is important to understand which instructions lead to which type of errors in the application software program when faults happen in the underlying hardware and how these faults are masked/propagated to higher system layers. In particular, this work develops novel techniques for cross-layer software program reliability modeling and optimization at different levels of granularity (e.g., instruction and function) and at different system design abstractions in order to compose and execute application software programs in a reliable fashion. Important highlights of the novel contributions of this manuscript are:

Cross-Layer Software Program Reliability Modeling and Estimation: This work develops cross-layer reliability analysis, modeling, and estimation concepts, techniques, and tools. An extensive program reliability analysis is performed to understand the manifestation of hardware-level faults at the software level. This analysis is leveraged to devise software-level reliability models that account for the hardware-level knowledge in order to bridge the gap between the hardware and software for accurate reliability estimation at the software level. At the instruction granularity, the following models are developed to capture key reliability aspects of a software program:

1. The *Instruction Vulnerability Index* estimates the probability of an instruction's output being erroneous due to soft errors. It accounts for *spatial vulnerabilities* (i.e., area-wise error probabilities) and *temporal vulnerabilities* (i.e., time-wise error probabilities) of different instructions executing in different pipeline stages of a given processor while accounting for hardware-level information, e.g., the probability of faults in different processor components obtained through a detailed gate-level soft error analysis.
2. The *Instruction Error Masking Index* estimates the probability that an error at an instruction will ultimately be masked until the final program output, i.e., does not become visible at the application output and therefore is denoted as "masked."
3. In case the error is not masked, the *Error Propagation Index* estimates how many outputs will be affected by the unmasked error.

These instruction-level estimates are then used to obtain the reliability estimates at basic block and function/task levels. In the optimization flow, these models are leveraged to quantify the reliability-wise importance of different instructions, basic blocks, and functions to enable selective reliability optimization at different system layers under tolerable performance overhead constraints.

Cross-Layer Software Program Reliability Optimization: This manuscript develops concepts and techniques for cross-layer reliability optimization and leverages multiple system layers for reliable composition and execution of application software programs. First multiple versions of a software program are obtained that enable run-time trade-offs between reliability and performance properties. This is done through the following two means:

1. *Different reliability-driven software transformations and instruction scheduling techniques* are proposed that lower spatial/temporal vulnerabilities and probabilities of software program failures and Incorrect Outputs by reducing the number of executions of *critical instructions* (like load, store, branches, jumps, and calls). Applying these transformations in constrained scenarios provides on average 60 % lower software program failures (i.e., crashes, halt, hang, abort) and thus increased software reliability.
2. *Reliability-driven selective instruction redundancy* is proposed that selects a set of reliability-wise important instructions in different functions for redundancy-based protection depending upon the instruction vulnerabilities, instruction-level error masking and propagation, and protection overhead under user-provided tolerable performance overhead constraint. The key is to give more protection to the less-resilient part of the software program and less protection to more-resilient part to achieve a high degree of reliability in constrained scenarios. Compared to state of the art, the proposed selective instruction protection provides 4.84× improved reliability at 50 % tolerable performance overhead constraint.

Afterwards, multiple reliable versions are exploited by a reliability-driven run-time system that enhances the reliability of multiple concurrently executing applications in a manycore processor while accounting for the frequency variations and degradation due to design-time process variation and run-time aging-induced effects. It performs the following key operations to facilitate reliable software program execution:

1. *Adaptively activating and deactivating the redundant multithreading for different applications* in a manycore processor in area-constrained scenarios. It accounts for variable resilience properties and deadline requirements of different applications along with a history of the encountered errors.
2. *Dynamically selecting an appropriate reliable version* for each application considering cores' frequency variations due to design-time process variations and run-time aging-induced performance degradation.
3. Mapping the selected application version on the cores used for redundant multithreading at run time such that the execution properties of the redundant threads closely match the frequency properties of allocated cores considering core-to-core frequency variations.

Compared to state-of-the-art single-layer reliability optimizing techniques, the proposed cross-layer approach achieves 16 %–57 % improved software reliability on average for different chip configurations, various process variation maps, and different aging years.

In addition to the above-discussed scientific contribution, several tools for gate-level soft error analysis, aging analysis, an integrated fault generation, and an injection system for instruction set simulators have been developed in the scope of this work and are made available at http://ces.itec.kit.edu/846.php.

Contents

List of Figures

List of Tables

List of Algorithms

Acronyms

AES	Advance encryption standard
AGU	Address generation unit
ALU	Arithmetic logic unit
ASIC	Application-specific integrated circuit
CAVLC	Context adaptive variable length coding
CDF	Cumulative density function
CDFG	Control and data flow graph
CEE	Common expression elimination
CPU	Central processing unit
CRAFT	Compiler-assisted fault tolerance
CRC	Cyclic redundancy check
DCT	Discrete cosine transform
DM	Data memory
DMR	Dual modular redundancy
ECC	Error correcting codes
EDDI	Error detection by duplicated instructions
EPI	Instruction error propagation index
f/MCycles	Faults per million cycles of execution
FPU	Floating point unit
FVI	Function vulnerability index
GB	Gigabyte
GCC	GNU compiler collection
H.264	The advanced video coding standard H.264 by MPEG and ITU
HT	Hadamard transformation
ID	Instruction decoder
IEU	Instruction execution unit
IF	Instruction fetch
IM	Instruction memory
IMI	Instruction error masking index
IPRED	Intraprediction
ISA	Instruction set architecture

ISS	Instruction set simulator
IVI	Instruction vulnerability index
IW	Instruction word
MB	Megabyte
MC-FIR	Motion-compensated FIR filtering
MCycle	Million cycles
NMOS	N-type metal-oxide-semiconductor logic
NPC	Next program counter
PC	Program counter
PMOS	P-type metal-oxide-semiconductor logic
RAM	Random access memory
RMT	Redundant multithreading
RTP	Reliability-timing penalty
SAD	Sum of absolute differences
SATD	Sum of absolute transformed differences
SDC	Silent data corruptions
SIMD	Single instruction multiple data
SRAM	Static random access memory
SWIFT	Software implemented fault tolerance
TCL	Tool command language
TMR	Triple modular redundancy
TSMC	Taiwan Semiconductor Manufacturing Company

Chapter 1
Introduction

Embedded computing systems are ubiquitous and have been widely deployed in many application domains like security, consumer, internet-of-things, mission-critical, and airborne applications. The current and emerging generations of embedded computing systems pose stringent design constraints in terms of performance, area, power, and cost. Furthermore, there has been an increasing trend of applications' functionality and even the number of applications on these computing systems. To cope with these trends and constraints, Moore's Law has paved the path of technological development for semiconductor industries, according to which the number of transistors doubles every 18 months [1]; see Fig. 1.1. Shrinking transistor features, like gate size and channel length, result in low power consumption, high performance, low cost per transistor, and high transistor density per chip. This miniaturization has facilitated the realization of powerful (embedded) computing architectures that provide high performance-per-power efficiency, which is evident from the processor roadmap, current generation of commercial processors with tens to hundreds of compute cores (like Intel Xeon Phi [3] and Nvidia GPUs [4]), and the International Technology Roadmap for Semiconductors (ITRS) [5]. However, as a consequence of this aggressive miniaturization, these computing devices fabricated with nano-scale transistors (i.e., beyond 65 or 40 nm) now face serious reliability threats (like soft errors, aging, thermal hot spots, and process variations) and robustness challenges at various abstraction levels. This challenges the feasibility and sustainability of further cost-effective technology scaling [5–8]. These reliability threats arise from multiple sources, as discussed below, and may result in faults in the hardware that have catastrophic effects on the correctness of applications' execution, thus raising inevitable challenges and concerns for the system designers, architects, and technology vendors [5, 9–12].

© Springer International Publishing Switzerland 2016
S. Rehman et al., *Reliable Software for Unreliable Hardware*,
DOI 10.1007/978-3-319-25772-3_1

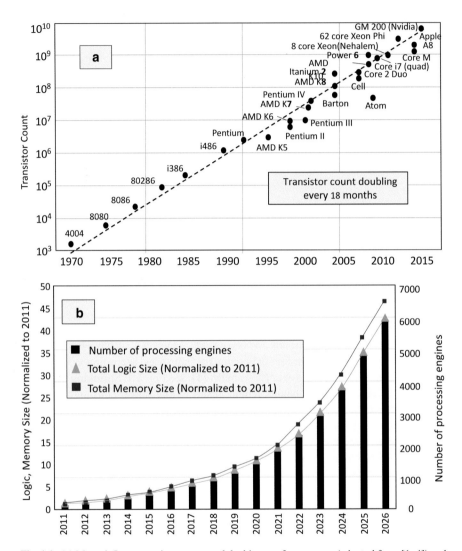

Fig. 1.1 (**a**) Moore's Law: transistor count and the history of processors (adapted from [1–4]) and (**b**) ITRS scaling trends for memory and processors [5]

1.1   Reliability Threats

The nano-scale transistors are subjected to various reliability threats that can be broadly categorized into the following classes.

Permanent and Intermittent Faults: There are two key sources: (1) manufacturing and structural faults like stuck-at faults (i.e., stuck-at-zero, stuck-at-one) and delay faults; and (2) aging faults, like Negative-Bias Temperature Instability (NBTI), Hot Carrier Injection, Time-Dependent Dielectric Breakdown, and

Electromigration. Aging faults manifest as delay/timing faults or frequency degradation if ensuring timing. Aging faults in the early years typically manifest as intermittent faults and in the later years as permanent faults. Other sources of intermittent faults are burst faults resulting from process variations, noise, voltage fluctuations, etc. The NBTI-induced aging is one of the most prominent aging threats, and it occurs in PMOS transistors when negative stress is applied at the gate [13]. An equivalent process called Positive-Bias Temperature Instability happens in NMOS transistors. The NBTI-induced aging manifests as an increase in the threshold voltage V_{th} by an amount ΔV_{th}, resulting in an increase in transistor delay leading to timing errors or a loss in the operating frequency over a period of time, typically in months to years. Considering a minimum operating frequency constraint for the lifetime of the chip, aging may result in a decrease in the lifetime if the delay degradation leads to a frequency drop beyond the required frequency level.

Process Variation is a design-time issue caused by the manufacturing-induced process variability. The variability challenge arises from the fact that nano-scale transistors are more difficult to precisely manufacture. Thus, the physical and electrical parameters of the transistors that come back from the foundry are different from the design intent. For instance, these transistors vary in terms of their doping, channel length, gate size, oxide thickness, transistor width, electron mobility, threshold voltage, etc. [14]. As a consequence of process variations, even identical transistors in two different locations on a chip can have different delay and leakage power properties, as can two transistors in the same location on different chips. Therefore, process variations result in variability in the operating frequency and leakage power consumption of processing cores on a chip (i.e., core-to-core variations, where all the cores comply to the same (micro)-architecture) and across different chips of the same or different wafers (i.e., chip-to-chip or die-to-die variations) [14, 15]. This can result in potential yield loss if a fraction of the chips does not meet the frequency and/or power specifications [16, 17, 25].

Transient Faults are manifested as single or multiple event upsets in the memory/sequential and combinational logics for single or multiple cycles [18]. Beside the electrical-noise, a major source of transient faults is *radiation-induced soft errors* that has emerged as one of the key reliability threats. Soft errors are transient faults due to low-energy or high-energy particle strikes[1] on the transistors [18]. If an energetic particle strikes the transistor, it creates charges (in the form of electron–hole pairs) that can switch on the transistor for a short time before the charges are diffused, thereby flipping the state of the logic element (i.e., logic gates) or the memory element (i.e., SRAM cells). These soft errors are manifested as *spurious bit flips* in the underlying hardware that may jeopardize the correct software execution. The number of bit flips and their duration depends upon the amount of charge produced by the particle strike (function of the incident particle's energy) and the physical/electrical properties of the transistor (affects the collection charge and the critical charge required to produce a bit flip). The bit flips in the memory elements

---

[1] Low-energy particles are typically alpha particles from the packaging materials while the high-energy particles are typically neutrons and protons from cosmic rays.

(e.g., pipeline registers, register files, and caches) are latched unless overwritten, while in logic elements could be of transient nature and may get masked due to subsequent gates [18, 19]. Although the low-energy particles can be avoided by using materials with low levels of impurity, the high-energy particles from the cosmic rays stay as a potential threat towards soft errors [18]. Due to the shrinking transistor dimensions and increased power densities, the critical charge to produce soft errors follows a decreasing trend, i.e., increased susceptibility to soft errors [19]. Intel predictions state that the bit Soft Error Rate will increase at a compounding rate of 8 % [20]. Although some studies still show that bit Soft Error Rate is almost constant or may even slightly decrease in FinFETs, due to the high integration density, the system Soft Error Rate is rapidly increasing as it is proportional to the number of transistors in the chip.

In summary, technology scaling in the nano-era leads to several reliability threats like soft errors, aging, and process variations that are expected to worsen significantly in the upcoming technology nodes [20, 21]. Furthermore, temperature serves as a catalyst for several reliability threats (like aging and soft errors). Therefore, reliability/dependability has emerged as one of the major design criteria in embedded computing systems. These reliability threats result in a variety of system errors ranging from Silent Data Corruptions to Application Program Failures (like crash, hang, and abort) that can have a catastrophic impact depending upon the application domain and the final system deployment [7, 22]. Industrial standards mainly rely on introducing excessive guardbands to mitigate aging and process variations [23, 24] in order to prolong the Mean-Time-To-Failure. The mainstream research efforts mainly explore *aging mitigation techniques* to lower these guardbands [23] or *exploit variability as an opportunity* to have a pool of architecturally similar cores with varying performance and power levels [25] or to *tolerate variation errors* [26]. Mitigating soft errors is, however, very challenging due to their random and transient nature [22] and most of the techniques are based on full-scale architectural redundancy [22] or program redundancy [27, 28]. Nonetheless, in order to achieve a cost-effective reliable design of modern embedded computing systems, all of the above-discussed reliability threats are important to be mitigated in one or the other way as they affect the system failure rates in all phases of chip lifetime.

This manuscript primarily targets soft errors, which are one of the most important reliability threats in the nano-era. Towards this end, this manuscript aims at leveraging multiple software layers to achieve high soft error resilience on unreliable or partially reliable hardware, while exploiting the inherent error masking characteristics and soft error mitigation potential at different software layers, which has not yet been explored. Since a real-world system is subjected to multiple reliability threats, *this manuscript will also present concepts to improve soft error resilience in the presence of aging and process variations.* However, mitigation of aging, temperature-dependent effects, and the variability problem is out of the scope of this manuscript. In the following, a major focus of discussion will be kept on the soft error problem, while aging and process variations will be discussed to the level of detail, which is necessary to understand the novel contributions of this manuscript related to soft error resilience under process variation and aging effects.

Before proceeding to the research challenges and novel contributions of this manuscript, increasing trends of soft errors are presented to highlight the importance of this issue.

1.2   Increasing Trends for Soft Errors

Soft errors have emerged as a non-negligible design challenge in hardware/software systems, and its importance is likely to increase with each upcoming technology generation [19]. The number of soft errors per unit time is termed as *Soft Error Rate* (SER). The unit of measuring the SER is Failure-In-Time (FIT). One FIT is equivalent to one failure in one billion device hours [18]. According to [18], the failure rates that are produced by soft errors are significantly higher than other reliability threats, i.e., up to 50,000 FIT/Chip compared to 50–200 FIT which is the collective failure rate that comes from other reliability threatening mechanisms, such as metal oxide breakdown and electromigration. The SER has accelerated with each upcoming technology node. Though for the newer technologies (like FinFETs), the trend of SER increase is different, prediction from technology vendors show that soft errors will still be an important issue and soft error rate will increase by a rate of 8 % per bit per technology generation as shown in Fig. 1.2 [29]. However, some studies also show that while the bit SER has saturated (i.e., either stays constant across different technologies or even slightly lower for the upcoming technologies), the system SER is increasing exponentially due to the high integration density, as discussed below.

Figure 1.3a presents the device scaling trends by Texas Instruments for the SRAMs that exhibit reduced cell (charge) collection efficiency. It is due to the fact that reduction in the cell depletion size is cancelled out by the lower operating voltage and reduced node capacitance [18]. Figure 1.3b shows that, although initially the SRAM single bit SER was increasing with every new technology node, in the deep submicron regime (<0.25 μm), the *SRAM bit SER has saturated* and may even be decreasing. This can be attributed to the saturation in voltage scaling, reductions in the junction collection efficiency, and increased charge sharing with the neighboring

Fig. 1.2 Intel's trends for soft error failures [10]

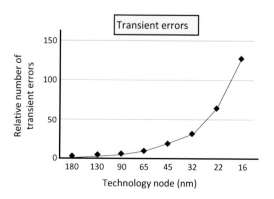

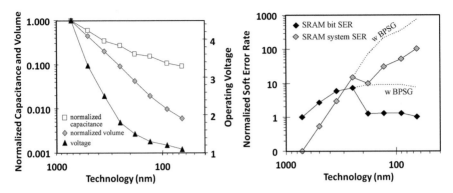

Fig. 1.3 (**a**) SRAM scaling and power trends and (**b**) bit and system soft error rates (SER) for SRAM [18]

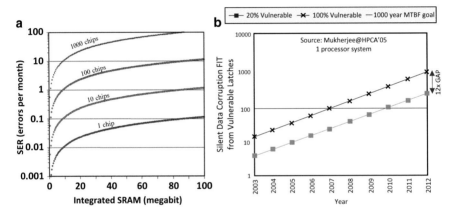

Fig. 1.4 (**a**) System soft error rates with increasing number of chips [18] and (**b**) silent data corruptions from vulnerable latches for different levels of protections [22]

nodes due to the reduced channel lengths. However, high integration density leads to an increased number of memory elements and logic gates on the chip, which results in an *exponential increase in the system SER* as it is proportional to the number of transistors on the die. Similar prediction trends can also be found in more recent studies by Ibe et al. [30].

Besides the complex microprocessor design, further increase in the system SER can also be attributed to the growing number of cores in an on-chip system and increasing trends of on-chip memory integration as shown in Fig. 1.4a. Besides SRAMs, due to the reduced critical charges and high integration density, there have also been increasing trends of soft errors in the combinational circuits, where soft errors in the gates can ultimately propagate to the output latch of the circuit [18]. Figure 1.4b shows the studies conducted by Mukherjee [22] that even after assuming 1000 years of Mean-Time-Between-Failures as adopted by IBM [31] and considering

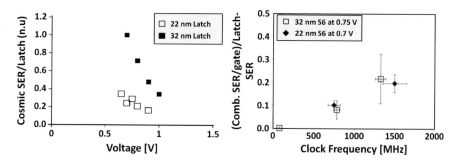

Fig. 1.5 (**a**) Voltage dependence of upset rates of 32 nm planar and 22 nm tri-gate latches. (**b**) Measured 32 and 22 nm combinational cosmic SER per logic gate expressed as a fraction of latch SER [32]

only 20 % vulnerable latches (i.e., 80 % latches are protected using ECC, a practical case of this is Fujitsu SPARC that had 20 % of 200k latches vulnerable), the lifetime constraint can no longer be met in the modern systems due to the increasing number of cores. However, in more cost-constrained scenarios, the number of vulnerable latches is very high. The gap between the required reliability level and Silent Data Corruptions is about 12x if 100 % latches are considered vulnerable, i.e., a pessimistic design option in cost-constrained scenarios. The gap may even exceed to 100x for an 8 processor system [22]. These studies showed that a data center with over 300 multi-core processors experiences a data corruption almost every week.

Soft Error Trends for FinFETs (i.e., Tri-Gate Technology): Figure 1.5 presents the SER for SRAM and combinational logic for 32 nm planar and 22 nm Tri-Gate (FinFET) devices [32]. It is shown that the FinFET devices have low susceptibility to soft errors compared to the planar CMOS devices, for instance, SRAMs using FinFETs at 22 nm yield an approximately 3.5x reduction in the SER compared to SRAMs fabricated using a 32 nm planar CMOS technology. The work in [32] performed scaling of 32 nm planar to 22 nm planar and speculated that SER for FinFETs at 22 nm is still less than that in the 22 nm planar devices. It is shown that the SER per logic gate as a fraction of the Latch SER is increasing at higher operating frequency, though the SER of 22 nm FinFETs is still less than the SER in 32 nm planar devices.

To address the non-negligible soft error issues, there have been extensive investigations to devise soft error mitigation techniques both at the hardware and at the software levels. State-of-the-art hardware-level reliability techniques are primarily relying on full-scale architectural redundancy (i.e., dual/triple modular redundancy and parity protection) [33] and often incur significant overhead in terms of area and power/energy. To alleviate this overhead or to complement the existing hardware-level techniques in (area/power)-constrained scenarios, several software-level reliability techniques have evolved, for instance, Error Detection using Duplicated Instructions [27] and Software Implemented Fault Tolerance [28]. These software-level techniques are based on data/code redundancy and control flow checking [34] and may also incur significant performance and memory overhead ($\geq$2x–3x), which is prohibitive within the stringent design constraints of embedded computing systems. The compiler-level techniques [72, 73] primarily target only the register file component (which is a small part of the processor [35]), and do not consider the complete processor perspective during reliability estimations and optimization, thus leading to only limited reliability improvements (about 2–7 %). Soft error resilience in multi-/manycore processors is mainly achieved by redundant multithreading that exploits idle cores to execute redundant threads and comparison/voting. Prominent examples of such works are Intel's Chip-level Redundant Threading [36], Process-Level Redundancy [37], and Reunion [38]. However, these techniques do not account for core-to-core frequency variation knowledge from the hardware layer during the soft error resilience. Consequently, these techniques may lead to *either* significant synchronization delays that can violate the performance constraints (i.e., degraded timing) of different critical tasks *or* excessive mismatches/rollbacks because the result of the redundant thread is delayed and not available at the same time instant. In general, a majority of hardware-/software-level techniques *do not exploit the soft error masking potential available at different software layers* and therefore, may lead to an overdesign, which may be cost-inefficient. Furthermore, most of the software-level reliability techniques lack knowledge from the hardware layers (i.e., where faults occur) and may lack reliability efficiency and fault coverage. *This instantiates the need for a cross-layer reliability solution where knowledge from different hardware and software layers must be leveraged to achieve cost-effective reliability in constrained scenarios.*

To summarize, traditionally, reliability is mainly addressed at the hardware level. However, due to the unavoidable hardware errors, designing cost-effective solutions to resolve these reliability issues is, nowadays, a task that must also be considered at software levels by exploiting inter-layer error masking properties, as will be discussed in Chaps. 3 and 4. State-of-the-art software-level reliability techniques have, by far, not exploited their potential since the common belief, so far, was that reliability problems when occurring at the hardware level should be addressed at the hardware level and such techniques have not explored the knowledge and masking potential available at different software and hardware layers in a synergistic way. Furthermore, in a real-world scenario the susceptibility of the on-chip multi-/manycore systems towards multiple reliability threats is inevitable, thus *joint consideration of aging and process variation-induced effects during the soft error resilience* is crucial. In short, to enable a highly reliable software system for embedded computing, it is crucial for system designers to *leverage multiple system layers* in an integrated fashion for joint optimization of reliability under constrained scenarios, i.e., given user-provided tolerable performance overheads. Designing such a multilayer software reliability system however, poses several research challenges in terms of modeling and estimation as discussed in the following section.

1.3 Research Challenges for Enabling Cross-Layer Software Reliability

One way to solve the reliability threats is at the lower layers (i.e., device/transistor level) from where these faults are stemming. However, handling these problems at the device level may require significant amount of design effort, verification and validation time, and cost that may jeopardize sustaining Moore's Law with completely fault-free transistors in a cost-effective way. Furthermore, as discussed above, a lot of transistor-level faults get masked at the architecture and software level, or may not even matter for the current execution context at a given time instance. Therefore, the semiconductor and hardware/software communities have recently experienced a shift towards mitigating these reliability threats at higher hardware/software design abstractions and exploiting their masking potentials, rather than completely mitigating these at the device level.

The software-level reliability optimization may even be more important in case the available processor lacks full-scale protection against soft errors (and potentially also against other reliability threats) or may even be unreliable. In such cases, the challenge is how to enhance the reliability at multiple software layers in order to reduce the software programs' susceptibility to soft errors and to improve the reliability of the overall system while accounting for the knowledge/information exchange across multiple system layers. It is even more challenging when considering the stringent design constraints for embedded computing systems, for instance, reliability optimization under tolerable performance overhead constraints. However, handling the hardware-level faults at the software-level is not straightforward and poses several research challenges regarding reliability modeling and optimization at the software layers, as outlined below.

Research Challenges for Enabling Accurate Reliability Estimation at the Software Level: To enable software-level reliability models, it is important to analyze the effects of hardware-level faults at the software level in terms of functional and timing correctness, the corresponding masking effects, and error manifestations (i.e., incorrect program output or program execution failures like crash and abort). Faults in different processor components during the execution of different instructions may result in different types of faults. Such a study is important to be made to devise effective models for quantifying software program's reliability. Traditional software-level reliability models ignore the hardware-level knowledge (like fault distributions and microarchitectural properties) that may lead to inaccurate reliability analysis and estimation at the application software level. Hence, *there is a need to bridge the gap between the hardware and software* by quantifying the effects of hardware-level faults at the software level, while jointly considering the processor knowledge (like layout/area, logical masking properties of different processor components that depends upon their microarchitecture, and fault distribution) and software properties (like error probabilities of different instructions in different pipeline stages, masking effects due to varying data and control flow). Furthermore, it is also important to *identify the appropriate granularity of software-level reliability models* to enable optimizations at different software layers.

Research Challenges for Enabling Cross-Layer Software Reliability Optimization: Software reliability optimization under constrained scenarios needs to account for the knowledge from different system layers, for instance, (1) resilience or error masking properties of different functions and even of different instructions within a given function; (2) space- and time-wise error probabilities of different instructions; (3) fault distributions and masking properties of the underlying hardware components; (4) knowledge about the hardware-level protection, e.g., instruction decoder and program counters are protected, but execution units and register files not; and so on. There exists a significant potential for reliability optimizations at the compiler and system software layers that has not yet been fully exploited due to the unavailability of the instruction-level reliability models. In order to employ *compiler-level reliability optimization*, it is important to quantify the reliability-wise importance of different instructions and identify a set of the most reliability-wise important instructions. Furthermore, such reliability estimation needs to be performed using static methods. Therefore, a key question that this manuscript aims at addressing is: *if and how can code generation be exploited to improve the reliability of the generated code of a given application software program executing on unreliable hardware under tolerable performance overhead constraints.*

Another related challenge is to determine the *granularity of applying the optimization methods,* i.e., whether reliability optimization at the instruction granularity is more beneficial or at the function granularity or even both can be employed considering different software layers. For instance, employing instruction-level reliability optimizations during reliable code generation, and function-/task-level reliability optimization during reliable program execution. *Such a cross-layer optimization flow needs to account for the reliability potential available from different layers.* For instance, different reliable code versions may be available for the system software to execute for the same function/task. In such cases, the reliability-driven system software needs to analyze the functional and timing reliability properties of such reliable code versions to select an overall reliability-optimizing solution for the current execution context.

Besides the above-mentioned challenges or as part of them, this manuscript targets the following research challenges at different software layers:

1. Characterizing the reliability-wise importance of different instructions and functions.
2. Enabling dependable code generation and dependable application program execution.
3. Enabling interactions between reliability-driven compilation and reliability-driven system software.

4. Enabling reliability-performance tradeoffs at compile and run time.
5. Improving the software error resilience in multi-/manycore processors under core-to-core frequency variations due to manufacturing-induced process variability and aging.
6. Jointly accounting for functional reliability (i.e., correctness of program's output in the presence of faults) and timing reliability (i.e., deadline misses in the presence of faults, for instance, due to reliable application execution with performance loss).

1.4 Contributions of This Manuscript

In order to address the above-discussed research challenges, this manuscript introduces a novel cross-layer approach where the key goal is to leverage multiple layers in the system design abstraction to exploit the available reliability enhancing potential at each system layer and to exchange this information across multiple system layers. In particular, this manuscript aims at reliable code generation and execution on an unreliable or partially reliable embedded hardware that is subjected to multiple reliability threats. To enable this, the following novel concepts and techniques for cross-layer reliability modeling and optimization under user-provided tolerable performance overhead constraints are proposed. Figure 1.6 illustrates an overview of the proposed novel contributions.

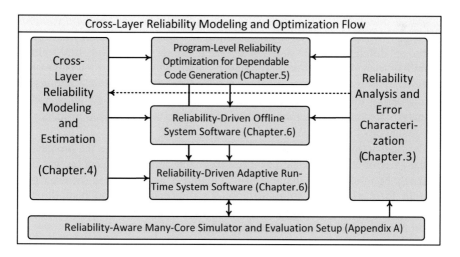

Fig. 1.6 Overview of the contributions of this manuscript

1.4.1 Cross-Layer Software Program Reliability Modeling and Estimation

Novel *cross-layer reliability analysis, modeling, and estimation* concepts, techniques, and tools are proposed in this manuscript that exploit the knowledge from both software and hardware layers to efficiently model the reliability factors at the corresponding system layers adapted to the appropriate granularity (e.g., instruction and function/task level). First, an extensive reliability analysis of application software programs is performed to understand (1) which instructions lead to which type of errors in the software programs when faults occur in different processor components; and (2) how different errors get masked at the software layers. This analysis is leveraged to develop cross-layer software reliability models for quantifying the effects of hardware-level faults at the software level, while accounting for the knowledge of the processor architecture, layout, and fault probabilities of different processor components, in order to bridge the gap between the hardware and software. This has not yet been considered by state-of-the-art software-level reliability modeling approaches [73–76].

The proposed software reliability models aim at quantifying error probability, error masking, and error propagation effects at different granularity to enable reliability optimization at the corresponding granularity in different system layers. In particular, this manuscript introduces the following novel models/metrics to quantify software reliability in the presence of hardware-level faults.

1. The **Instruction Vulnerability Index (IVI)** model estimates the probability of an instruction's output being erroneous due to soft errors. It jointly accounts for *spatial vulnerabilities* (i.e., area-wise error probabilities) and *temporal vulnerabilities* (i.e., time-wise error probabilities) of different instructions executing through different pipeline stages of a given processor. Quantifying these vulnerabilities for a given application software program requires: (a) spatial vulnerability analysis w.r.t. different pipeline components (like register file and ALU) considering their area and fault probabilities; and (b) temporal vulnerability analysis for different instructions that exhibit varying vulnerable residency in different processor components.

2. The **Instruction Error Masking Index (IMI)** model estimates the probability that an error at an instruction as a result of fault during its execution will ultimately be masked until the final visible output of an application software program. Estimating IMI requires error masking analysis depending upon the data flow properties (e.g., instruction type and value of operands variables) and control flow properties (e.g., execution probabilities of basic blocks).

3. The **Instruction Error Propagation Index (EPI)** model quantifies the error propagation effects at the instruction granularity, i.e., if an error at an instruction is not masked until the final visible output of the application software program, then to how many program outputs will it propagate and affect. This requires an analysis of non-masking error probabilities in multiple instruction paths in the data and control flow graph.

4. The **Function Vulnerability Index (FVI)** and **Function Resilience** models estimate the reliability at the function granularity. The FVI model estimates the function reliability as a function of IVIs of its instructions. However, the function resilience is a black-box model that estimates resilience of a function to hardware-level faults as a probabilistic measure of function's correctness. The function resilience model is beneficial in cases, where designers do not want to dive into in-depth program analysis.

5. The **Reliability-Timing Penalty (RTP)** model is beneficial in cases where besides functional correctness, timing reliability plays also an important role, e.g., in timing-conscious embedded systems. The *Reliability-Timing Penalty* is defined as a linear combination of the functional reliability penalty (e.g., quantified as FVI) and timing reliability penalty (e.g., quantified as probability of deadline misses).

These reliability models are leveraged by different techniques in a cross-layer reliability optimization flow for quantifying the reliability-wise importance of different instructions and functions/tasks and to perform constrained reliability optimization.

1.4.2 *Cross-Layer Software Program Reliability Optimization*

Novel *cross-layer software reliability optimization* concepts and techniques are proposed in this manuscript that leverage multiple system layers for dependable code generation, application composition, and application execution. Dependable code generation for a given application software program is enabled through applying different reliability-driven transformations, instruction scheduling, and selective instruction protection techniques in a *reliability-driven compilation* flow. Whereas, dependable application composition and execution is enabled via a reliability-driven system software that accounts for core-to-core frequency variations as a result of design-time process variations and run-time aging. It manages application execution in redundant multithreading mode while considering reliable code versions. The proposed cross-layer software reliability optimization aims at improving the reliability of application software programs through two important means: (1) reducing the error probabilities by lowering the spatial and temporal vulnerabilities of different instructions; and (2) error detection and recovery in constrained scenarios through selective instruction and task-level redundancy. The proposed techniques improve reliability under user-provided tolerable performance overhead constraints.

In particular, this manuscript introduces the following novel concepts and techniques for cross-layer reliability optimizations at different granularity at different system layers as summarized below.

1. The **Reliability-Driven Code Generation** optimizes the reliability of application software programs by employing several reliability-driven software transformations, instruction scheduling, and selective instruction redundancy

techniques under tolerable performance overhead constraints. The reliability-driven transformations and instruction scheduling technique aim at reducing error probabilities while selective instruction redundancy targets error detection and recovery support at the instruction granularity. These code generation techniques are leveraged to generate a set of different compiled versions for each application function, such that, different function versions differ in terms of their reliability and performance properties. A brief overview of the proposed novel concepts and techniques for dependable code generation is presented below.

(a) **Reliability-Driven Software Transformations**: The following four reliability-driven transformations are proposed that lower the error probabilities by reducing the instruction vulnerabilities and number of *critical instruction* executions.

- *Reliability-Driven Data Type Optimization* reduces the number of executions of *critical instructions* (like load, store, branches, calls, and address computing instructions) by transforming smaller bit-width data types into larger bit-width data types for given data structures, while minimizing the function vulnerabilities.
- *Reliability-Driven Loop Unrolling* determines an "appropriate" unrolling factor such that the spatial and temporal vulnerabilities of different instructions are jointly reduced under performance and code size constraints. It explores the tradeoff between reduced loop evaluation instructions and vulnerabilities of live register variables.
- *Reliability-Driven Common Expression Elimination and Operation Merging* reduces the vulnerabilities by removing identical expressions and/or merging partially common sub-expressions. It investigates the reliability effects of re-computation and register variables with increased lifetime while also accounting for the register spilling.
- *Reliability-Driven Online Table Value Computation* evaluates whether pre-computed table values with increased memory vulnerability would be beneficial from the overall function reliability perspective or the online computation with increased instruction vulnerabilities in the pipeline.

(b) **Reliability-Driven Instruction Scheduling** determines the instruction execution sequence that influences the vulnerabilities of different instructions in different processor components, for instance, the variable vulnerable residency in the register file or instruction's vulnerable residency in the pipeline stages. The proposed instruction scheduler prioritizes instructions with the highest *reliability weight*, which is a joint function of instructions' vulnerabilities, probabilities of different error types during their execution, and their dependent instructions. To account for the knowledge of the reliability weights of dependent instructions, a *lookahead-based heuristic* is employed.

(c) **Reliability-Driven Selective Instruction Protection** selects a set of reliability-wise important instructions in different functions for employing redundancy-based protection under a tolerable performance overhead. For achieving high reliability in constrained scenarios, the proposed technique prioritizes instructions based on their vulnerabilities, error masking and propagation effects, and the protection overhead. The key is to give more protection to the less-resilient function of an application software program and less protection to more-resilient functions, when considering a performance overhead constraint.

(d) **Multiple Reliable Compiled Function Versions**: The above-discussed code generation techniques are used to generate multiple versions for each function of an application software program. These functions are identical in terms of their functionality and final output, but differ in terms of their reliability and performance properties. This enables the realization of a *reliability-performance optimization space* that will be leveraged by the system software layers for dependable application composition and execution in constrained scenarios, as discussed below.

2. The **Reliability-Driven System Software** exploits the multiple reliable function versions for dependable application execution on single core and multi-/manycore processors. It aims at improving the soft error resilience while accounting for other reliability threats like aging and process variations. Two novel contributions are made at the system software layers.

(a) **Reliability-Driven Offline System Software** generates multiple offline function schedules for reliability-driven application composition. It constructs function schedule tables, such that each function schedule represents a particular dependable application composition using a different set of reliable function versions. The goal of the offline system software is to minimize the total *Reliability-Timing Penalty* of the complete function schedule. For this, it picks an appropriate reliable version for a function considering the reliability and execution time properties of the previously executed functions and the remaining time until the application deadline.

(b) **Reliability-Driven Run-Time System Software**: The *run-time system software for single core* processors dynamically selects an appropriate schedule from the offline generated function schedule tables depending upon the current execution context and the application deadline. In case of multi-/manycore processors, different cores may operate at different frequencies due to the design-time process variations and/or run-time aging-induced frequency degradation. Therefore, the *run-time system software for manycore* processors aims at achieving soft error resilience considering core-to-core frequency variations. It employs an adaptive *Dependability Tuning* technique that performs three operations for dependable application execution. (1) It adaptively activates/deactivates the redundant multithreading for different concurrently executing applications under resource constraints (e.g., available cores) while accounting for variable resilience properties of different applications, their deadlines, and a history of the encountered errors. (2) It performs variation-aware thread-to-core mapping for tasks with and without redundant multithreading to ensure both functional and timing reliability. The key idea is to find a set of cores for tasks in the redundant multithreading mode such that the redundant threads finish execution at almost the same time before the deadline, thus reducing the synchronization overhead and avoiding deadline misses. (3) It selects an appropriate reliable function version depending upon the frequency of the allocated core and the deadline.

Note that these software-level techniques may be employed in conjunction with hardware-level techniques to further improve the system reliability.

1.5   Orientation of This Work in the SPP 1500 Priority Research Program on Dependable Embedded Systems

The work in this manuscript contributes towards the subproject *GetSURE— Generating and Executing Dependable Application Software on UnReliable Embedded Systems* as part of the German Science Foundation (DFG)-funded priority research program *SPP 1500—Dependable Embedded Systems* [156]. The goal of SPP 1500 is to provide improved error resilience and online adaptivity in order to address various dependability threats, like soft error and aging, which arise from the fabrication imperfection, transistor miniaturization, thermal hot spots, and other side effects of the continuous technology scaling; see Fig. 1.7. The aim of this priority program is to devise novel modeling and optimization techniques across various layers in the system design abstraction ranging from circuits all the way up to the application level in order to improve the reliability of the overall hardware/software system. One of the key goals of this SPP is to enable support for cross-layer reliability to achieve cost-effective reliability by exploiting the knowledge from multiple hardware and software layers. *This manuscript covers the cross-layer software reliability aspects within this SPP while considering the knowledge from the underlying hardware layers as an input.*

Figures 1.7 and 1.8 show different research facets of the SPP 1500 program highlighting the involvement of different system layers like hardware architecture and software subsystem covering design, monitoring, and management aspects.

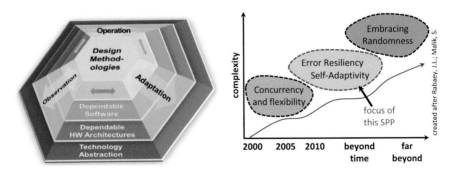

Fig. 1.7 Pyramid for dependable embedded systems and focus of SPP 1500 [156]

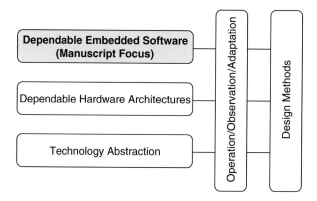

Fig. 1.8 Different research facets of SPP 1500 and the orientation of this work [156]

This SPP aims at enhancing the dependability through various means, amongst many, few are discussed below.

1. *Technology Abstraction:* The aim is to provide a clean interface between the technology-related research and higher abstraction layers in order to facilitate independent yet effective research of novel techniques/methods at multiple hardware and software layers.

2. *Dependable Hardware Architectures:* This ranges from the lower logic to the microarchitecture and architectural-level hardware and all the way up to the system-on-chip hardware architectures. Other subprojects like OTERA [153, 154] aim at achieving high fault tolerance through reconfiguration and self-organization.

3. *Dependable Embedded Software*: This covers research and development of novel concepts and methodologies for generating reliable code, managing reliable execution of application code on different architectures, and analyzing the impact of different hardware-level faults on the software level. A key goal is to enable soft error tolerance of the embedded software (including both the application program and operating system) to the reliability threats emerging in the hardware and manifesting at the software layers.

4. *Operation, Observation, and Adaptation:* Overall dependability of an on-chip system is a joint function of the reliable hardware architecture and reliable embedded software, which cannot be viewed in an isolated way. Therefore, efficient mechanisms for observation of erroneous scenarios and adaptation of different components of the hardware/software subsystems are important to achieve cost-effective dependability in unpredictable scenarios. Key goals are to explore techniques like error recovery and failure avoidance.

This work makes contributions to the following columns of SPP 1500: (1) Dependable Embedded Software; and (2) Operation, Observation, and Adaptation; see filled box in Fig. 1.8. In particular, this manuscript aims at cross-layer reliability modeling and optimization techniques to enable dependable code generation and execution on unreliable hardware. Both offline and online software techniques are proposed in this manuscript in case the underlying hardware components are unreliable or partially protected. In order to bridge the gap between hardware and software techniques for achieving a high degree of reliability, this manuscript will perform a detailed characterization of the impact of hardware-level faults at the software layer and will leverage this program reliability analysis to develop various reliability models at the instruction and function granularity. These models will be employed to guide different software layers for reliability optimization.

1.6   Outline of the Manuscript

The manuscript is organized in eight chapters, where Chap. 1 introduces the reliability problems, challenges, and novel contributions. Chapter 2 provides the necessary background on different reliability threats and state-of-the-art techniques in reliability modeling and optimization at both hardware and software levels. Chapter 3 provides a comprehensive overview of the proposed cross-layer reliability modeling and optimization flow along with a detailed program reliability analysis using fault injection. Chapters 4, 5, and 6 provide detailed techniques, algorithms, and evaluation of novel contributions of this manuscript. Results for the cross-layer optimization are presented in Chap. 7. Chapter 8 concludes the manuscript and presents some visions on potential future works. A brief outline of each chapter is given below.

Chapter 2: Background and Related Work: This chapter discusses background knowledge on different reliability threats, for instance, the preliminaries on soft errors, aging, and process variations, their sources and basic mechanisms (see Sects. 2.1, 2.2, and 2.3). Section 2.4 will present state-of-the-art soft error analysis, modeling, and estimation techniques at different hardware/software layers highlighting their limitations and benefits. Afterwards, hardware-level soft error mitigation techniques are discussed in Section 2.5.1 along with their benefits and associated overheads and limitations. To alleviate the overhead of different hardware-level techniques, related works have devised several software-level soft error mitigation techniques that will be discussed in Section 2.5.2 highlighting their basic limitations. The ignorance of hardware-level information renders these techniques as inefficient.

Chapter 3: Cross-Layer Reliability Analysis, Modeling, and Optimization: This chapter presents the system overview of the proposed cross-layer reliability analysis, modeling, and optimization flow in Sect. 3.1, highlighting the novel modeling and optimization techniques at different system layers and their interactions. The novel concept of bridging the gap between the hardware level (where the faults occur) and the software level (where the reliability is aimed to be improved) is presented, while highlighting its importance for efficient reliability modeling and optimization. A detailed program-level error analysis is performed in Sect. 3.2 using fault injection experiments and an output error characterization is presented. A characterization of instructions w.r.t. their relationship to the output error types is performed in Sects. 3.2.1 and 3.2.3. This analysis also helps in understanding the effects of hardware-level faults at the software-level and thereby identifying different parameters for cross-layer software reliability modeling. Afterwards, a high-level overview of the proposed cross-layer reliability modeling flow is presented in Sect. 3.3 highlighting how different parameters from both the hardware and software levels are considered to devise multiple software reliability models at different levels of granularity. The detailed techniques and results of these models are discussed in Chap. 4. Towards the end, a novel cross-layer software reliability optimization flow is presented in Sects. 3.5 and 3.6 that employs several novel techniques at different system layers for dependable code generation, application composition, and execution. The detailed techniques and results of these reliability-optimizing techniques are discussed in Chaps. 5 and 6.

Chapter 4: Software Program-Level Reliability Modeling and Estimation: This chapter presents the details on the proposed cross-layer software reliability modeling and estimation that are developed at different granularity, i.e., instruction and function-/task-level. First, the instruction-level models will be presented, namely, Instruction Vulnerability Index (Sect. 4.1), Instruction Error Masking Index (Sect. 4.2), and Instruction Error Propagation Index (Sect. 4.3) that quantify the error vulnerabilities, error masking probabilities, and error propagation effects at the instruction granularity, respectively. Besides discussing the parameter estimation and information from both hardware and software levels, results for these models will be presented for different applications and their variations will be discussed w.r.t. the diversity in the instruction profiles. Afterwards, two different function-level reliability models, namely Function Vulnerability Index and Function Resilience, will be introduced in Sect. 4.4. In Sect. 4.4.1, the Function Vulnerability Index will be discussed in detail as it will be used for the reliability-optimizing techniques proposed in this manuscript. Function resilience provides an alternate modeling solution and will be discussed in Appendix B. Afterwards, the Reliability-Timing Penalty model will be presented in Sect. 4.4.2 that provides a joint function for functional and timing reliability and will be beneficial for timing-conscious embedded systems.

Chapter 5: Software Program-Level Reliability Optimization for Dependable Code Generation: This chapter presents several novel techniques for dependable code generation to improve the application software program's reliability under user-provided tolerable performance overhead constraint. First, four novel reliability-driven software transformations are presented in detail in Sect. 5.1 along with concept explanation, examples, and algorithms. For each transformation, its impact on the software reliability is analyzed. Afterwards, detailed error distribution results for different versions generated after applying these transformations are presented and analyzed. Based on these results, selection of appropriate transformations for different applications is discussed. Furthermore, the impact of these transformations on the critical address generation arithmetic instructions, overhead reduction for instruction redundancy techniques, and function vulnerability reduction is analyzed. In Sect. 5.2, the reliability-driven instruction scheduling technique is explained in detail along with examples, cost function, and algorithm discussions. The proposed technique is compared with state-of-the-art instruction schedulers that ignore the instructions' vulnerabilities from the full processor perspective. The comparison is made for different error types after extensive fault injection campaigns. Moreover, reduction in function vulnerability is also analyzed and discussed. Section 5.3 will present the reliability-driven selective instruction protection technique with the help of an example explaining the concept, detailed algorithm and cost function, and comparison results with state-of-the-art highlighting the benefits of considering the proposed reliability models for this optimization. Towards the end, Sect. 5.4 will present the concept of multiple reliable function versions and selection of versions lying on the pareto-frontier. These function versions enable the reliability-performance optimization space for the reliability-driven system software of Chap. 6.

Chapter 6: Dependable Code Execution using Reliability-Driven System Software: This chapter presents the techniques to enable dependable application composition and execution using the system software layers that exploit the concept of multiple reliable function versions. First, a reliability-driven offline system software is discussed in Sect. 6.1 that constructs reliable function schedules considering different execution contexts. Examples and algorithm details are presented along with the Reliability-Timing Penalty evaluation of the proposed techniques. Afterwards, *reliability-driven function scheduling for single core processors* is presented in Sect. 6.2 that selects an appropriate reliable function schedule at run time based on the deadline and execution properties of the previously executed functions. Finally, Sect. 6.3 presents the *reliability-driven system software for manycore processors* that employs a novel *Dependability Tuning technique* to improve the soft error resilience in the presence of core-to-core frequency variations due to process variations and aging. Since the proposed *Dependability Tuning* technique integrates the novel contributions of this manuscript, a comprehensive evaluation of it is performed in Chap. 7 as it will also demonstrate the benefits of cross-layer software reliability optimization compared to single-layer reliability optimization techniques in constrained scenarios.

Chapter 7: Results and Discussions: This chapter presents the results and benefits of the proposed cross-layer reliability approach when compared with different single-layer reliability-optimizing techniques. First, the processor synthesis, aging estimation, and process variation maps are presented in Sect. 7.1. Section 7.2 briefly presents the benchmark applications from the MiBench suite that are evaluated for reliability and for fault injection campaigns. A summary of comparison results over different chip sizes, numerous process variation maps, and various application scenarios are presented for different aging years in Sect. 7.4. Afterwards, detailed comparison results are presented in Sects. 7.5 and 7.6 for different chip sizes, and some selected chips, respectively.

Chapter 8: Summary and Conclusions: This chapter presents a short summary of the manuscript.

Appendix A: Simulation Infrastructure: This appendix presents the integrated tool flow and simulation setup which is developed under the scope of this work. First, the reliability-aware manycore simulator is discussed which is based on ISA-simulators for cores with SPARC v8 instruction set architecture generated using the ArchC tool chain. The simulator has integrated fault generation and injection modules that are discussed along with different parameters.

Appendix B: Function-Level Resilience Modeling: This appendix presents the function resilience modeling, which is a black-box model to estimate the degree of output correctness for a given function.

Appendix C: Algorithms: This appendix presents detailed algorithms for different modeling and optimization techniques proposed in this manuscript.

Appendix D: Notations and Glossary: This appendix presents a table of notations/symbols used in this manuscript and their definitions/descriptions. Furthermore, a glossary of important terms is also presented.

Chapter 2
Background and Related Work

This chapter presents the background knowledge regarding different sources of the emerging reliability threats (i.e., soft errors, process variation, and aging-induced effects), the related work on soft error modeling, and their mitigation techniques. In particular, Sect. 2.1 provides the background regarding soft errors, starting with the basic transistor structure and its functionality, followed by various soft error sources and the soft error mechanism. Section 2.2 presents the basics of the NBTI-induced aging phenomena. Section 2.3 presents different variability sources and manufacturing induced process variation effects along with the process variation model explained in Sect. 2.3.1. Section 2.4 discusses the related work on soft error modeling and estimation at both the hardware and software layers. Starting from the traditional to more advanced approaches, Sect. 2.5 presents state-of-the-art soft error mitigation techniques at both hardware and software levels. As the focus of this work is on soft errors, most of the background discussed is related to soft errors. Towards the end, Sect. 2.6 summarizes the related work.

2.1 Soft Error

2.1.1 Transistor Structure

Before going into the details of how soft errors becomes an issue in the transistors, it is important to have a basic knowledge of the transistor's structure and functionality. The fundamental unit of the CMOS (Complementary Metal-Oxide-Semiconductor) microprocessor's underlying structure is the *n-type* (NMOS) and *p-type* (PMOS) transistors. The n-type transistor carries the electrons, whereas the p-type transistor carries the holes from source to drain. Figure 2.1 shows the structure of the NMOS transistor. A metal gate is attached to a thin layer of a silicon dioxide (SiO_2) which forms the interface with the silicon substrate. Channel formation is necessary for the

© Springer International Publishing Switzerland 2016
S. Rehman et al., *Reliable Software for Unreliable Hardware*,
DOI 10.1007/978-3-319-25772-3_2

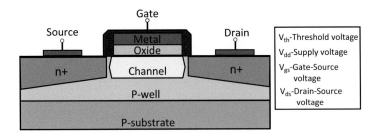

Fig. 2.1 Structure of an NMOS transistor

flow of mobile charge carriers (i.e., electrons/holes) through the transistor. A channel will be formed from source to drain and the transistor will be considered as ON, when the gate-source voltage V_{gs} becomes larger than the threshold voltage V_{th}. Otherwise, the transistor remains OFF as no channel is formed and no electrons/ holes are travelling from source to drain.

2.1.2 Soft Errors in Transistors

Soft errors may be caused due to *external events* like energetic particle strikes, and/ or *internal disruptive events* like noise transients at circuit, chip or system level, cross talks, and electromagnetic interference [39]. As compared to alpha particles and low energy thermal neutrons, the high-energy cosmic neutrons produce huge amount of energy (e.g., in the range of 80 MeV–1 GeV) when it strikes the nuclei of a silicon substrate in the chip [18, 135]. Figure 2.2 illustrates the soft error mechanism that can be explained in the form of the following three main phases [18].

1. **Phase-1: Ion-Track Formation**: When an energetic particle (like high-energy neutron from cosmic rays) strikes the semiconductor material (e.g., substrate of a transistor), the energy transfer results in the generation of numerous electron–hole pairs and a high carrier concentration along the ion's path. This is called the *ion track*. The high-energy cosmic neutron produces 80 MeV–1 GeV of energy with a single strike and every 3.6 eV of energy the ion loses, produces one electron–hole pair.

2. **Phase-2: Ion Drift and Current Pulse Generation**: The produced charge is collected within a few microns at the junction. In this ion drift phase, when the ions come close to the depletion region, the electric field collects the carriers that results in a generation of a current spike (or voltage transient). This process is called "funneling." Collection of charge near the depletion region can result in a temporary formation of a channel (even if the transistor is originally in the OFF state) and consequently leads to the flow of electrons from source to drain. This sudden current glitch may result in an instantaneous power-on of the transistor for a very short period of time (typically for tens of picoseconds).

3. **Phase-3: Ion Diffusion**: Over the period of time, the ions are diffused in the transistor for instance in the depletion region. This illustrates the transient nature of

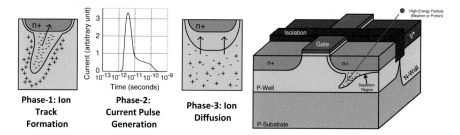

Fig. 2.2 Soft error mechanism illustrating different phases of charge generation, collection, and diffusion

soft errors. In this process, additional charge is collected until all excess carriers are captured, recombined, or diffused away.

With sufficient amount of collected charge, bit flips may result in logic devices and get latched into memory elements. These bit flips may corrupt the state of the processor (i.e., content in pipeline registers, register files, caches) and jeopardize the correct application software program execution. With the miniaturization of transistors, close spacing, and reduced critical charges, increasing trends of multi-bit upset (even from a single particle strike) have recently been reported in the literature [84, 85].

Critical Charge and Collection Charge: There are two important factors related to the soft error mechanism: (1) collection charge $Q_{\text{Collection}}$ and (2) critical charge Q_{Critical}. As transistor dimensions reduce, the critical charge also decreases along with the operating and threshold voltages. The $Q_{\text{Collection}}$ is the amount of charges (i.e., electrons/holes) collected in the conducting path of the transistor (i.e., between source and drain) to form a channel. Whereas, the Q_{Critical} is the specified number of accumulated charges which are required to form a channel between source and drain. When the $Q_{\text{Collection}} \geq Q_{\text{Critical}}$, the channel can form and the electron/holes start flowing from source to drain. Now as the Q_{Critical} reduces, a lesser number of accumulated charge is required to form a channel between the source and drain [135]. Thus, the chances for experiencing a soft error become higher, because a higher number of electron/hole pairs are generated upon a particle strike that can rapidly accumulate to form a channel in the conducting path between the source and drain. In short: lower Q_{Critical} means, fewer number of collected electrons/holes can form a strong channel. Furthermore, the Q_{Critical} becomes lower with higher temperature and further aggravates the soft error rate [117]. When an energetic particle strikes the silicon substrate, the $Q_{\text{Collection}}$ is modeled using Eq. 2.1:

$$Q_{collection} = Q_{all} \exp\left(-\frac{x_c}{L_{max}}\right) \tag{2.1}$$

Typically, the $Q_{\text{Collection}}$ is in the range of 1 to several 100 fC, and its exact amount depends upon the type of the energetic particle, its path, and the dissipated energy along the path. Likewise, from the transistor perspective there are several factors upon which the amount of $Q_{\text{Collection}}$ is dependent, i.e., size, substrate structure, location of the strike, and device state. The transient current pulse for ion-track charge collection typically has a double exponential form with rapid rise time and gradual fall time as shown below in Eq. 2.2 [105].

$$I(t) = \frac{Q_{collection}}{\tau_\alpha - \tau_\beta} \left(e^{-\frac{t}{\tau_\alpha}} - e^{-\frac{t}{\tau_\beta}} \right) \tag{2.2}$$

The parameters τ_α and τ_β denote the collection time constant and the time constant for the ion-track formation, respectively. Both these parameters are dependent upon the process with typical values for $\tau_\alpha = 164$ ps, and $\tau_\beta = 50$ ps. The $Q_{Critical}$ denotes the amount of the critical charge required to change the data state. Its value can be expressed as Eq. 2.3 [86].

$$Q_{critical} = \int_0^{T_F} I_D(t)\, dt \tag{2.3}$$

The parameter I_D denotes the time-dependent drain transient current and T_F denotes the flipping time, which is the time when both the voltages at the drain and at the gate become same [86]. The above model works well for simple circuits like DRAM storage cells. However, more complex circuits like SRAM cells experience an upset when the recovery time of the cell τ_r (time taken for the struck node voltage to return to its pre-strike value) exceeds the feedback time τ_f (time taken for the struck node voltage to become latched as incorrect data) [106]. Therefore, for the computation of $Q_{Critical}$, $T_F = \tau_r$ if the cell recovers and $T_F = \tau_f$ if the cell upsets. In general, $T_F = \min(\tau_r, \tau_f)$.

2.1.3   Masking Sources for Soft Errors

In combinational circuits, not all soft errors in the underlying hardware propagate to the output due to different masking effects. Masking means the ability of a logic circuit to prevent soft errors from occurring/appearing at the final output. Figure 2.3 illustrates the three major masking effects namely *Logical Masking*, *Electrical Masking*, and *Latch-Window Masking*.

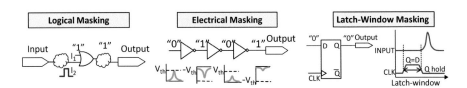

Fig. 2.3 Soft error masking effects [87]

The **logical masking** effect is defined as the capability of a logic circuit to prevent soft errors from affecting the final output. For instance, in the figure it is shown that the output of the OR gate will always be "1," as long as I_1 maintains a logic high state ("1"). The blue pulse is caused upon a radiation event, and this will not affect the final output. In the second figure, there is an electric pulse caused by the radiation event. However, the amplitude of the pulse is not strong enough to trigger a bit flip on the input signal. The pulse will be attenuated when it passes through each gate and finally dies out. This phenomenon is called **electrical masking**. In the third figure, there is a D-latch. The value of output Q depends on the clock, if the clock is high, the value of D is set to Q, if the clock is low, Q will hold its previous value. The latch window is defined as "this" interval, during which the clock is high. If the pulse misses the window, it will not affect the final output, due to the so called **latch-window masking**.

However, logic circuits still face a lot of soft error threats in spite of the masking effects, and many state-of-the art reliability estimation and optimization techniques are proposed at both the hardware and at the software levels (discussed in Sects. 2.4 and 2.5) to protect the logic circuit from soft errors or make sure that even if a soft error happens, the circuit can detect the errors and give the correct output.

2.2 NBTI-Induced Aging

There are different mechanisms for aging like Negative-Bias Temperature Instability (NBTI), Hot Carrier Injection, and Time-Dependent Dielectric Breakdown. NBTI-induced aging has emerged as one of the most crucial aging phenomena that happens in the PMOS transistor. An equivalent process called Positive-Bias Temperature Instability happens in NMOS.

Figure 2.4a illustrates the NBTI mechanism in a PMOS transistor which is under stress, i.e., the gate voltage is minus V_{dd}. When this voltage is applied, it creates a force at the inversion layer resulting in the breakdown of the silicon and hydrogen bond at the interface of the silicon and oxide layer (as negative stress attracts the positive hydrogen ion). The hydrogen ion is released in the oxide layer and at the broken bond a trap is created which can trap any free ions or charges, thus making the insulation imperfect. Even two neutral H atoms can combine together into an H_2 molecule, which can escape from the surface of the oxide [65]. At a higher abstraction layer, the NBTI-induced effect is manifested as an increase/shift in the threshold voltage, thereby making the transistor slower. Note, the exact phenomenon is still not precisely known and is an actively researched area in device physics [143].

The temperature plays an important role in further accelerating the NBTI-induced aging effects, i.e., increase in the threshold voltage shift. Figure 2.4b shows different curves for the threshold voltage shifts over a period of time for different operating temperatures [19]. As the temperature increases, the shift in threshold voltage aggravates and therefore increases the delay, and thus the aging effects become more prominent [151]. When estimating the change in the threshold voltages for two temperatures,

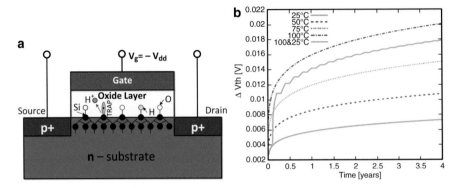

Fig. 2.4 (**a**) NBTI-induced aging (adapted from [137]) and (**b**) Impact of temperature on the NBTI-aging [19]

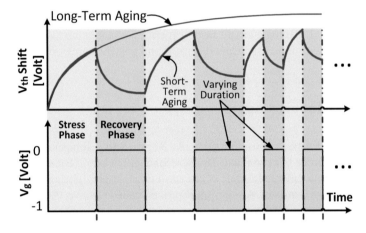

Fig. 2.5 Short-term and long-term aging [138]

i.e., 75 and 50 °C, the ΔV_{th} is approximately 50 % higher at 75 °C than 50 °C. It can be seen that the NBTI effects for two alternating temperatures, i.e., between 100 and 25 °C, are worse than that at 75 °C [151]. This shows that the shift in threshold voltage ΔV_{th} will be determined by the higher temperature.

As soon as the negative stress is removed, a recovery phase is triggered but 100 % recovery happens only in infinite time. Figure 2.5 shows an abstract view of the stress phase (causing V_{th} increase) and recovery phase (causing V_{th} decrease) along with the short-term and long-term aging effects. The stress and recovery phase can be explained using the *reaction–diffusion model* [40]. One reason for the partial recovery could be that the hydrogen ion again tries to re-bond with the silicon, but this behavior is not fully understood [142]. However, 100 % recovery is not possible as re-bonding is a random process and the same hydrogen and silicon atoms will never always perfectly re-bond. It is reported that the recovery is probably better at higher temperatures. However, higher temperatures also aggravate the threshold voltage shift in the stress phase.

If aging is not properly taken care of, then the cores' safe operating frequency (i.e., to ensure correct execution) and system clock frequency become different and may lead to timing errors. To compensate the increase in the threshold voltage V_{th} by an amount ΔV_{th}, the circuit needs to execute at a lower frequency by a factor of Δf that may violate the performance constraints, otherwise the circuit output may be faulty due to the timing errors. The industrial practice to solve this issue is via guard banding, i.e., running all the cores at the slowest frequency, which will lead to a system-wide performance loss. The aging-induced performance/delay degradation varies depending upon the stress produced due to workload and operating conditions [19]. During the first year, the aging of the core is expedited and dependent upon the core usage, whereas in the years onwards the long-term aging happens which is in times of months and years and is more dependent on the temperature [19, 138]. The device-level NBTI aging model (Eq. 2.4) employed in this work is obtained together with the *VirTherm-3D* group [151] as a part of the collaborative research effort in the DFG SPP1500 program [118]. It is based on the reaction–diffusion theory [13, 40].

$$\Delta V_{th} = 0.05 \times e^{-1500/T} \times V_{dd}^4 \times y^{1/6} \times d^{1/6} \tag{2.4}$$

ΔV_{th} is the mean threshold voltage shift in volts, T is the temperature in kelvin, V_{dd} is the supply voltage in volts, y is the age of the transistor in years, and d is the duty cycle, i.e., probability that the transistor is stressed.

Note: the aging model adopted in this work is based on the reaction–diffusion theory but another aging model based on trapping–detrapping theory [143] can also be employed because the proposed algorithms and concepts are orthogonal to the aging models. The aging values (in form of core-to-core frequency degradation) serve only as an input to the proposed cross-layer reliability modeling and optimization flow for evaluating concepts related to soft error resilience under frequency variations.

2.3 Manufacturing-Induced Process Variations and Other Variability Sources

The magnitude of the *process variations* (e.g., in the channel length/geometry and random dopant fluctuations) increases with the scaling technology trends, as it is more difficult to precisely manufacture smaller transistors with exactly the specified dimensions [25, 45, 139]. One source of variability comes from the manufacturing side which manifests in the form of core-to-core frequency variations. Figure 2.6 shows the frequency variations in the Intel's 65 nm 80-core test chip, where at 1.2 V the maximum core frequency is 7.3 GHz and the minimum is 5.7 GHz. This corresponds to 25 % frequency variation on the same chip, whereas across different chips the variation will be more [42]. The cores running at different clock frequencies may lead to timing errors. To address this issue, the major industry practice is to do guard banding by running every core with the minimum frequency on the chip, i.e.,

Fig. 2.6 Frequency
variation in an 80-core
processor within a single
die in Intel's 65 nm
technology [42]

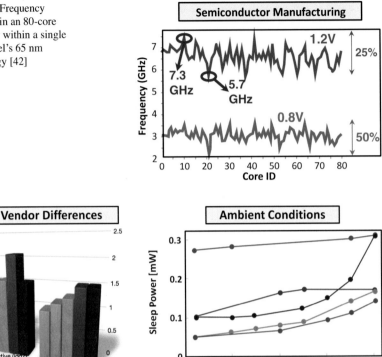

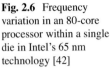

Fig. 2.7 (a) Power variation across five 512 MB DDR2-533 DRAM parts and (b) variation in sleep power (P_{sleep}) with temperature across five instances of an ARM Cortex M3 processor [26, 43]

in this case it is 5.7 GHz which is the slowest of all the cores. In synchronous system design, the core's performance is determined by the slowest critical path, and process variations may introduce severe *design-time performance degradation.*

Another source of variability comes from different vendors. This is because different vendors use different design rules and cell libraries to fabricate the same specification, ultimately leading to variations in the leakage and dynamic power across different chips. Figure 2.7a presents the variations in the maximum active and max idle power for five different vendors fabricating the same standard, i.e., DDR2 533 DRAM chips. The standard is fixed which is the DDR2 533 but as the vendors are different there are variations in the maximum active/dynamic power and maximum idle/leakage power across different vendors.

There are also ambient conditions-dependent variabilities. This comes with the variations in the temperature, e.g., as the temperature is increased the leakage power also increases. In a single 80-core chip, different cores might have different temperatures, resulting in different leakage power and performance properties. Figure 2.7b shows the measurement of the leakage power for different temperature for different variants of five ARM cortex M3 processors. It can be seen that at the

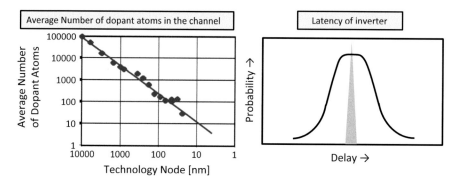

Fig. 2.8 Design-time process variation [144]

same temperature, different processors will have different leakage powers. Furthermore, a single processor shows variations in the leakage power as the temperature varies. This can be explained due to the fact that at higher temperatures, the electron mobility increases making it easier for them to escape from source to drain or from source to substrate, causing current leaks.

Figure 2.8 shows the trend of design-time variability that comes from random dopant fluctuations in the transistor. Doping is done in silicon semiconductors by implanting additional doped atoms, i.e., p+ and n+ in the transistor substrate. In the nano-scale transistors, it becomes increasingly difficult to manufacture every transistor with precisely the same number of dopant atoms, and with exactly the same dimensions, i.e., gate thickness and channel length. Earlier in one micron technology, there were more than 5000 of these dopant atoms which basically meant that minor variations of 4–5 atoms (0.001 %) had negligible effect on the electrical properties of the transistor. However, in the recent 32 nm technology, the number of dopant atoms is approximately 40 [19], and slight variations in the number of dopant atoms will significantly affect the properties of the transistor, e.g., a variation of four dopant atoms will become 10 % of the total that may result in performance degradation by approximately 10 %. The variations at the transistor level will ultimately reflect at the gate level, e.g., see the delay distribution for an inverter as shown in Fig. 2.8. In earlier technologies, the inverter was having a deterministic delay value which meant that all the inverters on the chip had exactly the same delay, e.g., 1 ps. But now due to process variations, inverters made from different transistors typically have different delay properties. For instance, some transistors may have a delay value of 1 ps while some others have 1.1 or 0.9 ps.

2.3.1   Process Variation Model

This manuscript employs the process variation model, which is proposed in [25]. It models the chip surface as a fine grid of dimensions $N_{chip} \times N_{chip}$. The process parameter value p_{ij} ($i; j \in [1; N_{chip}]$) at a grid cell (i, j) can be modeled as a Gaussian random variable with mean μ_p and standard deviation σ_p. The correlation between the process parameters at two different grid points is given as the correlation coefficient $\rho_{ij,kl}$ that reduces with increasing distance. Based on the experimentally validated model of [44], $\rho_{ij,kl}$ is given as Eq. 2.5.

$$\rho_{ij,kl} = e^{-\alpha\sqrt{(i-k)^2 + (j-l)^2}}, \quad \forall i, j, k, l \in \left[1, N_{chip}\right] \qquad (2.5)$$

The parameter α denotes the reduction rate of $\rho_{ij,kl}$. The frequency of a digital circuit can be modeled as the worst-case delay of the N_{cp} identical critical paths. According to [25, 45], the maximum frequency of core C_i ($i \in [1; N]$) in a multi-/manycore processor is modeled using Eq. 2.6, where K' is a technology-dependent constant and $S_{CP,i}$ denotes the set of N_P grid cells in C_i

$$f_i^{MAX} = K' \min_{k,l \in S_{CP,i}} \left(\frac{1}{\rho_{kl}}\right) \qquad (2.6)$$

As discussed in Chap. 1, since the primary focus of this work is on the soft error related issues, in the following sections, related work for the soft error modeling and mitigation techniques will be discussed in more detail.

2.4 State-of-the-Art Soft Error Estimation Techniques

In literature, extensive research has been conducted for analyzing and modeling the soft error impacts at various granularities, i.e., at circuit-/architecture level [47, 88, 93, 94] and program level [73, 74, 76]. The standard techniques are either based upon fault injection simulations or analytical/mathematical models developed to estimate the soft error propagation across multiple gates in the combinatorial circuits. In the following, an overview of these approaches at different design abstraction layers is given.

2.4.1 Circuit-Level Techniques

In circuits, the fundamental entity in both the logic and memory parts is the NMOS and PMOS transistor which are prone to soft errors and its sensitivity grows when fabricated in scaled technology nodes [10]. The soft errors are of major concern in memories because of their large footprint in the chips. Furthermore, the soft error-induced bit flips inside memories stay unless overwritten. To handle soft errors in memories, ECC-based techniques are prominent [62, 63]. In contrast, the soft error effects in the logic part of the combinatorial circuits are relatively less frequent due to various circuit-level error masking effects (i.e., logical, electrical, and latch-window masking [87]) that prevent the errors to appear at the final output in several cases. However, despite these error masking effects, the soft error failure rate in the combinational logic is becoming more crucial for transistors fabricated in the scaled technology nodes. This is primarily due to their reduced sizes and critical charges which makes them more prone to soft errors [21]. Moreover, achieving soft error detection and recovery in a cost-effective way is relatively difficult in combinational circuits compared to that in memories (typically protected with parity or ECC) because of the random and *transient* nature of these faults and a high degree of propagation to multiple memory elements. In general, the soft error rate measurement in circuits is in accordance with the potential error masking effects. A model is developed in [21] to measure the Soft Error Rate (SER) in the microprocessors while taking into account the effects of electrical and latch-window masking effects. At first, the SER is computed via simulating the mechanism when a particle strikes the gate till the drain of the gate. This behavior is simulated using charge to voltage pulse modes, and then the electrical model is used to check the characteristics of the voltage pulse reaching the latch input. Afterwards, for checking the pulse strength at the latch input, a pulse latching model is developed that checks the amplitude and duration to cause a soft error.

IBM developed the Soft-Error Monte Carlo Modeling program to check if the chip designs meet SER specifications [46]. Work in [47] only focuses on the electrical masking effects and has developed a mathematical model to analyze the soft error propagation across multiple gates in a combinational circuit. In [88], an analysis and modeling approach has been proposed to measure the SER while taking into account the error masking effect. Low-level HSPICE simulation is performed to obtain the electrical masking computation for each path, and logical masking computation is carried out by flipping the logic value at each input vector and each path independently. Works in [48, 49] present a reliability evaluation, where in [49], different error masking factors are separately computed for investigating the soft error tolerance of the circuit. In [48], probabilistic transfer matrices are used where each gate is represented as a matrix and for each input combination, the probability of its output value is explicitly known. However, the work presented in [48] focuses only on the logical masking effect of the circuit for given gate output probabilities without considering electrical and latch-window masking.

In [95], fault injection techniques are utilized which are based upon Monte-Carlo simulations which are time consuming because numerous experiments are performed to achieve certain accuracy. In [50], an electrical masking model is proposed in which soft error rate estimation is performed at chip level while taking into account the impact of voltage fluctuations, and it is reported that ignoring the voltage fluctuation in electrical masking can lead to inaccurate estimates of the soft error rate. Furthermore, the chances of experiencing both single and multiple bit flips are higher in the recent technologies along with the possibilities of multiple correlated bit flips. Hence it is important to account for such a correlation during soft error rate estimation, as proposed in [21]. In [51], a circuit-level technique is proposed which is based upon the error propagation probability. This technique uses a path-based analysis to check the error propagation from source to the outputs, which is a very useful and fast technique for reasonably accurate identification of the vulnerable parts of the design. In [52], a hybrid technique is proposed to compute the soft error vulnerability of the entire microprocessor system consisting of regular and irregular structures. The techniques at the architectural and logic level are integrated for estimating the soft error vulnerability of regular (address-based structures, i.e., register file, cache) and irregular structures (i.e., logic, functional units) within a microprocessor. In general, the circuit-level techniques cannot account for the soft error masking factors at the higher system layers like architecture and software program. Therefore, a large body of research has investigated soft error estimation techniques at both architecture and software program levels.

2.4.2 Architecture-Level Techniques

The *Architectural Vulnerability Factor* (AVF) model is developed in [31] which is employed by different state-of-the-art to estimate the soft error impacts at the architectural level. The AVF of a processor component is the probability that a fault in that component will result in a visible error in its final output. It is the fraction of faults that can appear at the output in the form of user visible errors. For AVF estimation of a processor component, the *Architecturally Correct Execution* (ACE) analysis [31] is performed for all the bits in the processor component. ACE bits are the bits which are deemed necessary for architecturally correct execution, meaning that any fault in these bits will affect the correct program output when no error correction techniques are employed. All other bits are termed as un-ACE bits, because a fault in these bits will not cause a user visible erroneous program output. A bit is said to be un-ACE for a fraction of time if a fault at its value does not affect the final output of the program otherwise it is ACE. In case of storage cells, the AVF is the percentage of the time that it holds ACE bits, whereas in case of logic structures, the AVF is the fraction of time (percentage of total execution cycles) that the ACE bits or instructions are processed. It is assumed that all bits are ACE unless proven un-ACE and this may lead to overestimating the vulnerability of the target processor component. It is reported in [89] that ACE analysis overestimates the vulnerability up to 7*x*. Furthermore, the circuit-level masking factors such as electrical and latch-window masking in a hardware component cannot be ignored during the soft error rate analysis of a hardware component. However, the ACE analysis ignores error masking, making this approach inappropriate for *irregular hardware structures* (e.g., functional units) and only suitable for *regular structures* such as cache, register file, and reorder buffer in microprocessors [90]. It is reported that different processor components and microarchitectures have distinct AVF values, e.g., a fault in an ALU might affect the final output, whereas a fault in branch predictors might incur performance penalty but will leave no impact to the final output. The work in [73] proposed the Register Vulnerability Factor (RVF) model which considers the register bits required for architecturally correct execution. Extending the concept of AVF, the RVF models consider the fact that soft errors in the register file can be overwritten and will have no impact on the final program output if read after being written. RVF is the probability that a soft error in registers can be propagated to other processor components (i.e., functional units, memory). While AVF concepts focus on the effect of soft error propagation, the RVF presents the probability of soft error propagation to other hardware components. The authors in [91] quantified the impact of transient faults on the Alpha 21264 microprocessor by estimating the fault masking and identifying the vulnerable portions of the processor. Enhancements of the AVF are discussed in [31, 89, 90]. The AVF is primarily used to make decisions between two reliability implementations of a processor component, but cannot be used to compare different programs at instruction and function granularity for a given architecture. A program-level reliability model is required to consider the program properties, i.e., instructions, control flow and data flow, and program-level error masking effects. A program designer also requires an error characterization from the program's perspective. Furthermore, hardware-level reliability analysis and estimation techniques [31, 91, 96] require significant development time and a long experimental duration. To address these limitations, software program-level techniques [27, 28, 75, 77, 80, 81, 92] can be used.

2.4.3   Software Program-Level Techniques

Several software program reliability estimation techniques have been proposed which are analogous to AVF. In [76], an Instruction Vulnerability Factor model is developed to assess the criticality of the instruction through fault injection experiments but this technique lacks in-depth knowledge of vulnerable bits, time-wise error probabilities, and explicit quantification of program-level masking effects. Due to the coverage issues of fault injection (especially under varying inputs), the Instruction Vulnerability Factor inherently limits accuracy of soft error analysis and estimation. Vilas et al. in [74] introduced the concept of Program Vulnerability Factor (PVF) as a microarchitecture-independent metric. PVF is fundamentally an adaptation of the AVF by shifting the Architecturally Correct Execution (ACE) analysis from microarchitecture to the program level, i.e., in a software resource (e.g., compiler-visible architectural registers). Besides PVF's inaccuracy due to ignorance of the underlying hardware properties (i.e., the layer where fault happens), PVF's consideration of only the number of ISA-visible ACE bits does not provide a comprehensive knowledge of the program reliability, and thus further limits its accuracy. The reliability of a program also depends upon the type of instructions, its data/control flow properties, and temporal effects as discussed in Chap. 3. For a same number of ACE bits, a fault in one program might cause Incorrect Output or no effect (i.e., correct output), but in another program it might cause a program failure (e.g., crash) due to the use of a different instruction. Moreover, a particle strike at a certain location in the processor may manifest as a different error compared to a strike in other parts. ACE analysis by PVF ignores different types of the manifested error that may incur different reliability cost. PVF, however, considers all bits as ACE unless proven un-ACE which might lead to an overestimate of the program vulnerability. However, not all bits are of same vulnerability as some bits are more important for the correct execution and some might be less important. For example, faults in some bits may result in a crash and faults in some other bits may lead to data corruption, which is within the tolerable limit of the program user. A case of the tolerable limit of the program user has been demonstrated by the authors in [97] where the most significant bits are more vulnerable compared to the least significant bits. Since not all ACE bits lead to the same type of errors with the same intensity, program reliability analysis without the characterization of the manifested errors is incomplete. This instantiates the need for error characterization at the program level, given faults are injected in the underlying hardware at varying rates. Overall, the state-of-the-art approaches [31, 74, 75, 91] did not analyze the effects of changing fault rates on the program behavior when estimating the program reliability for a given system scenario. Furthermore, these reliability estimation models do not consider the knowledge of hardware-specific details, like chip footprint with processor details (e.g., area of different components, number of physical registers, fault probabilities of different processor components) for fault distribution and fault injection under different fault rates. Moreover, these techniques do not consider the time-wise error probabilities of different instructions in different components of a pipeline. Therefore, these software program-level techniques are based on abstract fault models and they will lead to over- or underestimation of the reliability.

2.4.4 Fault Injection Methodologies

Engineers typically employ fault injection to analyze and estimate the system reliability [96]. The "saboteur" or "mutant" [100] techniques are based on modifying the VHDL code. These VHDL-based techniques require precise processor models and timing details, thus requiring significant development time. Therefore, these techniques cannot be deployed in the initial design phases for the application designer. High-level simulator-based techniques typically produce the errors at the program layer in the early design phases. The technique of [101] uses a command-based injection of single-bit faults (from a fault database) in ASICs using its SystemC model. Authors in [99] propose a fault injection technique for digital signal processors. SymPLFIED is a program-level fault injection and error detection framework [98]. It enumerates transient faults in registers, memory, and computation blocks of hardware. However, it does not consider the knowledge of chip footprint in its machine model, which may lead to an inaccurate program reliability analysis. Moreover, due to its complex model evaluation it suffers from long experimental duration. The performance and analysis accuracy comparison is made against SymPLFIED in Appendix A.

2.5 State-of-the-Art Soft Error Mitigation Techniques

In order to mitigate the soft error effects, extensive research to develop reliability improvement techniques has been conducted at both the hardware level and the software level. In the following, some prominent hardware- and software-level techniques are discussed.

2.5.1 Hardware-Level Soft Error Mitigation Techniques

The hardware-level soft error mitigation techniques are tackled at the device level by adopting specialized process technology (e.g., using SOI process [55]) and materials during fabrication, at the circuit level by adopting specialized radiation hardened cells or redundant logic, and at the architecture level through redundancy in time or in space.

Device-Level Techniques: A major practice has been on exploring the possibilities of mitigating the soft error at the transistor/device level since it appears at the lower layers. The transistor-level solutions are primarily relying on the process technology, i.e., the way in which transistors are manufactured such that it becomes shielded against the radiation events like alpha particles and neutron strikes. Shielding against soft error means that the amount of collected charge $Q_{collection}$ at a transistor node once exposed to a radiation event is reduced, thus minimizing the chances of soft errors. Note, to prevent the soft error event it is important that the $Q_{Collection} < Q_{Critical}$. Adopting specialized fabrication processes and usages of some special material for fabricating soft error immune transistors is very effective, but it has a substantial overhead (in terms of, for instance, area and cost) when deployed throughout the processor. Moreover, the validation and verification costs of such approaches are considerable [19, 71, 116].

To overcome the soft errors due to the alpha particles, the semiconductor industry adopted various shielding techniques against the alpha particle-induced soft errors, e.g., by deploying thick polyimide (100 μm as stated in [53]) as an alpha particle protection layer because of their efficient thermal and electrical characteristics. With shielding, the alpha particle-originated SER is reduced to around 20 % [54]. However, such shielding solutions are typically not adopted in case of high-energy neutron-induced soft errors, because for shielding against neutron strikes, the thickness of the protection layer is required to be a minimum of approximately 10 ft in concrete, which is not feasible for almost all computing devices. As the neutron-induced SER is becoming increasingly common in the current technology, the shielding solutions do not completely eradicate the susceptibility of the device against all sources of soft error. Furthermore, deploying these techniques even for reducing the alpha particle-induced soft error is unaffordable because of the strict cost constraints.

Silicon-On-Insulator (SOI) has evolved as a promising technology for MOS/CMOS fabrications to protect against soft errors and in this way, has become superior to the conventional bulk CMOS process technology. In SOI technology, a thin-film layered insulator called buried oxide or silicon-insulator (see Fig. 2.9a) is placed in the substrate, instead of the conventional silicon substrate [55]. The transistor with SOI technology has the capability to reduce the $Q_{collection}$ upon the radiation event, because the silicon-insulator layer keeps the bulk silicon isolated, thus preventing the excess charge in the bulk silicon (induced by the radiation event) from propagating towards the source/drain, or device channel. It is reported by IBM that usage of the SOI process technology enables $5x$ reductions in SER for SRAM [56]. Although, the SOI technology appears to be an attractive option for reducing the SER and also in low power applications [102], nevertheless, their high manufacturing costs have made them less famous in the products that have strict design constraints. Another, well-known device-level solution is the Triple well (TW) process technique [57] which is different from the conventional CMOS process where twin-well transistor is constructed. This process technology alleviates the device sensitivity towards single event upsets. A TW device has a buried n-well layer ("deep n-well") that separates the p-well from the p-substrate; see Fig. 2.9b. The idea of burying the n-well layer is to collect the electrons generated in the p-well region of the NMOS device upon a particle strike before they are collected at the surface of the NMOS (source–drain channel). The gathered electrons below the p-well and at the deep n-well junction do not have an impact on the device state.

Circuit-Level Techniques: At the circuit level, the soft error problem is addressed either by deploying the radiation hardened cells [66], changing the device parameters [58], or introducing redundant circuits [59]. The radiation hardened cells are in practice and are prevailing in the electronic devices deployed in the space and military missions. Changing or tuning the device parameters may help in reducing the soft error rate, e.g., increasing the supply voltage which makes the $Q_{Critical}$ higher, hence the SER becomes lower [58]. The introduction of redundant or error detection/correction circuits into the target design has been well explored in order to recover from soft errors. A prominent example is the RAZOR approach that introduces shadow flip flops in the pipeline to recover from errors. The recovery is accomplished using a global clock gating of the pipeline and an error detection through shadow flip flops that receive a delayed clock. Figure 2.10 shows (a) organization of a processor pipeline with RAZOR flip flops (FF) after each pipeline stage, and (b) the timing of the

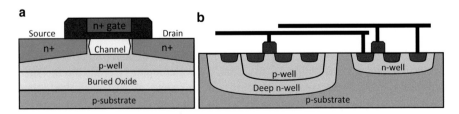

Fig. 2.9 (a) SOI MOSFET device and (b) TW NMOS FET structure

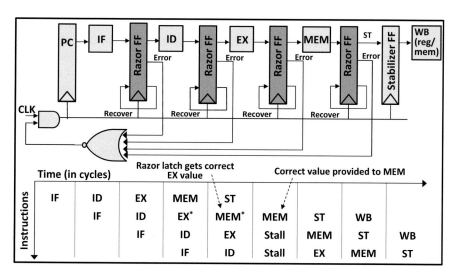

Fig. 2.10 A simple 5-stage pipelined processor with razor flip flop and error recovery [59]

overhead costs, these techniques are more practical and beneficial for the space-based applications. Furthermore, ECC and parity techniques are used to protect memories and caches. ECC-protected caches are a well-established practice in various research and industrial projects like IBM [62], AMD [63], and [35]. However, in case of register files, ECC is avoided due to high area and power overhead under frequent usage scenarios [27, 28, 73, 76], and consequently they remain vulnerable to soft errors. An approach to reduce this overhead by using the unused bits of registers is proposed in [152] but it has limited optimization potential in case of applications using full register widths and due to low soft error susceptibility of register file due to its very small footprint compared to the full pipeline. An alternate solution is employing parity-protected register files, but the error coverage is low, especially in case of multi-bit faults. Note, the contributions proposed in this manuscript are applicable to both protected and unprotected register files. Although the architecture-level solutions may not have the same precision and accuracy that the device-level solutions can offer, their efficiency is high because these solutions are independent from the underlying hardware-level details, i.e., process technology and transistor/cell structure. It is reported in [22] that when compared to the circuit-level hardening techniques, the ECC-based techniques [64] incur low area overhead. Besides the ASIC-based systems, techniques for improving the reliability of reconfigurable architectures have been proposed in [153, 154].

Redundant Multithreading Techniques: Multi-/manycore architectures facilitate soft error tolerance through excessive core availability, i.e., the cores originally reserved for performance improvements can now be exploited to improve the reliability through spatial and temporal redundancy [38, 83]. The works like fault detection via lock stepping [82], Simultaneous Redundant Threading (SRT) [67], Chip-level Redundant Threading (CRT) [36] and other works like [37] have focused pipeline for an error that happens in the execute (EX) stage (asterisk denotes an error in the pipeline stage computation). The detailed structure of the RAZOR flip flop is shown in Fig. 2.11 that employs a shadow latch controlled by a delayed clock. Deploying additional hardware structures, however, makes the circuit-level solutions more costly due to the incurred overheads in terms of area/power and verification cost, that may become prohibitive especially in the embedded systems.

Architecture-Level Techniques: At the architectural level, the availability of different functional units with distinct structures and diverse functional and timing properties makes a wider design problem when compared to device/circuit levels. The techniques at this level are based upon the redundant executions either in *space* (using the duplicated functional units) or in *time* (using the same hardware multiple times for redundant execution and comparing the outputs). Furthermore, keeping the redundant information can also help in recovering, for instance, from the corrupted state in memory. The traditional architectural redundancy approaches are Dual Modular Redundancy (DMR), Triple Modular Redundancy (TMR), Error Correcting Code (ECC), and parity protection. The DMR approach shown in Fig. 2.12 is used for error detection where two hardware modules are used to execute the redundant copy of a code and after the execution, a comparator checks the output. In case of a mismatch, the error is detected and the rollback execution is performed for recovery.

Figure 2.12 shows TMR which is used for error detection and recovery and that employs three hardware modules for executing three copies of the code. After the execution is finished, a majority voter compares the final outputs and selects the best two out of three to determine the correct output. Besides the large area/power overhead, the voter in TMR is the single point of failure. To overcome this, triplicated voters are deployed. The increased power overhead of TMR systems may potentially increase the temperature. Increased temperatures lead to higher SER due to the reduction in the critical charge [117] and increased aging. Due to their high

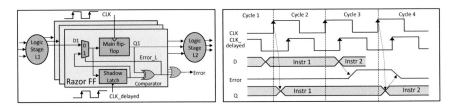

Fig. 2.11 Razor flip flop and timing diagram [59, 60]

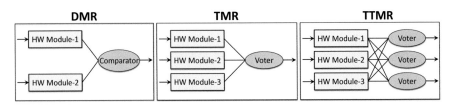

Fig. 2.12 Dual, triple, and triplicated triple modular redundancy [33, 36, 104]

on multi-/manycore based soft error mitigation where free cores are exploited to provide redundancy either at the hardware level (using redundant instructions or redundant threads) or operating system level (using redundant thread processes). Figure 2.13 presents fault detection via replicated microprocessors that are cycle-by-cycle lockstepped. This means that both the processors are synchronized with each other and have identical states at any point in time. At the same time, both processors receive same inputs and deliver the output at the same time. If an error happens in one processor, then the difference amongst the processor states will be detected and an error will be identified by the system monitor upon the output mismatch. The SRT [67] approach adapts the philosophy of the Simultaneous Multithreading (SMT) [68] approach that was originally proposed to improve the performance via executing the program codes of different applications in a simultaneous multithreaded fashion on multiple functional units inside a given processor; see Fig. 2.14. Instead of executing different application threads, SRT executes two redundant threads of the same application on multiple functional units (e.g., two adders, multipliers) inside the same core and then performs the output comparison. In contrast, the CRT approach executes redundant threads on two different processor cores; see Fig. 2.15.

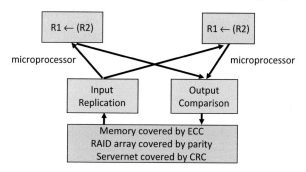

Fig. 2.13 Fault detection via lockstepping (HP Himalaya) [36]

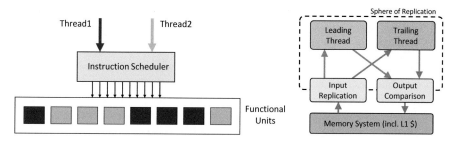

Fig. 2.14 Fault detection via simultaneous multithreading. *Left*: Scheduling different instructions on different functional units and *right*: sphere of replication with input and output replication [36, 67]

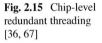

Fig. 2.15 Chip-level redundant threading [36, 67]

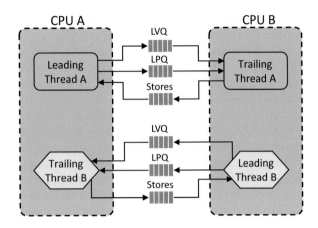

The SRT approach has an advantage of lower time-to-market and cost, as it exploits the existing well-established SMT architecture with little extra hardware. Moreover, it offers better performance than the complete replication. However, the challenge is the careful fetch/schedule of the redundant threads in a lockstepped fashion. The SRT combines both space- and time-wise redundancy. However, it prefers the space-wise redundancy due to its better coverage of permanent/long-duration faults. The CRT approach combines the best of SRT and lockstepping. Besides the conventional redundancy-based approaches, recent trends have evolved to explore flexible approaches, i.e., adaptive control of TMR/DMR [103], cores with heterogeneous error recovery functionalities [83], and the hardware-level checkpointing and recovery approaches [140].

The aforementioned approaches primarily target soft errors and may not address aging and process variation related problems during the soft error mitigation. For example, in CRT, two processor cores are executing redundant threads. In the presence of performance variations, one core may produce the output later than the other core. Hence, both the outputs may not synchronize for the comparison that will eventually lead to output errors. Alternatively, providing a large synchronization time may lead to performance degradation and potential deadline misses. Another limitation of these techniques is that they assume excessive area is available in the multi-/manycore system. However, in case of area-constrained embedded systems, there may be scenarios where not all applications may be supported with full DMR or TMR due to resource competition.

Summary: Within the scope of the hardware-based approaches, the soft error mitigation techniques have been explored at different abstractions in the system layers, i.e., device, circuit, and architectural level. Because of the prevailing facts that these hardware-level techniques have more area, therefore more power overhead and also high verification/validation costs, reliable hardware design and development using these techniques is both expensive and time consuming [19, 33, 71, 107, 116]. As a result, various soft error mitigation techniques at the software level have evolved. These software-level approaches are developed in various design abstraction layers which are hereby discussed in the following.

2.5.2   Software-Level Soft Error Mitigation Techniques

The classic soft error mitigation techniques (amongst many others) are N-version programming, code redundancy, control flow checking, and checkpoint recovery. The N-version programming [69] relies on implementing multiple program versions of the same specification. Depending upon the memory requirements, the number of program versions varies; however, N should not be less than 2. These N program versions are functionally identical but have diverse implementations leading to different failure characteristics such that not all versions fail in the same way under a given fault scenario. Figure 2.16 shows the working of N-version programming in a TMR model. For managing the execution, this technique demands a mechanism to synchronize the three outputs and to compare their results.

The software-based checkpoint recovery techniques [70] do not require a modification in the hardware and thereby restrict the area/power overhead. However, modification in the software is required in terms of additional implementations for the functionality. Such techniques place checkpoint instructions inside the code, typically before the critical instructions which have a high error probability. The program state is saved in reliable storage during normal execution (data checkpointing), so that in case of errors, the program can be resumed from the last checkpointed state. In [145], a compiler-assisted checkpointing scheme is proposed that inserts additional checkpointing code into user programs. An adaptive scheme is used to identify potential checkpoints in order to amortize the storage overhead of checkpointed data. The Libckpt [146] and libFT [147] checkpointing libraries provide routines to enable applications to dump critical data and/or states but require user intervention for maximum benefits. Efforts in [148] propose a reliable microkernel for application checkpointing that utilizes OS support to guarantee consistency between the current system state and process image. Besides checkpointing, a large body of work has been conducted at the software-/compiler-level that can be categorized in two major classes, (1) redundancy-based and control flow protection techniques and (2)

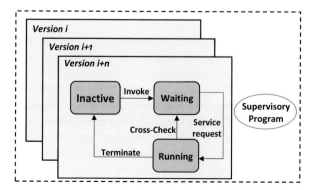

Fig. 2.16 N-version programming

Original code	Transformed Code [EDDI]
ld r12 = [GLOBAL]	ld r12 = [GLOBAL]
	1: ld r22 = [GLOBAL+offset]
add r11 = r12, r13	add r11 = r12, r13
	2: add r21 = r22,r23
	3: cmp.neq.unc p1,p0 = r11,r21
	4: cmp neq.or p1,p0 =r12,r22
	5: (p1) br faultDetected
st m[r11] = r12	st m[r11] = r12
	6: st m[r21+offset] = r22

Fig. 2.17 An example for EDDI [27]

vulnerability reduction-based techniques, as discussed below. These techniques offer new opportunities for constrained optimizations such that the cost budgets remain intact by exploiting the application characteristics.

The state-of-the-art redundancy-based techniques, such as Error Detection using Duplicated Instructions (EDDI) [27] and Software Implemented Fault Tolerance (SWIFT) [28], provide software reliability by duplicating the instructions, and inserting the comparison and checking instructions before the *store* and/or *conditional branches*. As a result, these techniques incur a significant performance overhead. An example is presented in Fig. 2.17 to explain the EDDI approach. In this example, the *load* from a global constant address is duplicated as instruction 1. In order to avoid conflicts between the original and duplicate instructions, the duplicated *load* reads its data from a different source address and stores its result into a different register. In a similar way, the *add* instruction is also duplicated as instruction 2 in order to create a redundant chain of computation. Finally, the *store* instruction is a point of synchronization, and instructions 3 and 4 compare the *store*'s operands (the address and the computed data) with their redundant copies. In case a difference is detected, instruction 5 will report an error. Otherwise, the original and its duplicate *store* instruction 6, will execute storing values to non-conflicting addresses.

As demonstrated in Fig. 2.17, this technique incurs a significant performance and memory overhead due to redundant instruction execution and shadow memory locations to store redundant data, respectively. Furthermore, performance overhead can also be attributed to the increased cache usage to hold redundant data for computation of original and duplicated instructions, generating additional memory traffic. With EDDI, although the input operands for branch instructions are verified, there is the possibility that a program's control flow gets erroneously misdirected without detection. The corruption can happen during the execution of the branch or register corruption after branch check instructions.

A well-established control flow technique is signature-based protection [34]. In order to verify that the control transfer is in the appropriate or intended basic

block, each block will be assigned a signature [34]. For that, a designated general
purpose register named as GSR (General Signature Register) is employed that
holds these signatures which are later used to detect faults. The GSR holds the
signature value for the currently executing block. As soon as there is an entry to
any block, the GSR will be *xor*'ed with a statically determined constant in order
to transform the previous block's signature into the current block's signature.
Once it is done the value inside the GSR can be compared with the statically
assigned signature for the block to ensure that an authorized control transfer has
occurred. In cases where two basic blocks have a control flow to a common block,
both the blocks can jump to a common block (a control flow merge) while sharing
the same signature. In such cases, using a statically determined constant to transform the GSR from the previous basic block signature to the current basic block
signature might not cover control flow errors. With the statically determined constant, faults which transfer control to *or* from blocks having the same signature
will remain undetected; this is undesirable. In order to avoid this, the signatures
should be determined dynamically.

Figure 2.18a highlights this technique, where instruction 1 and 2 are the redundant duplicates for the *add* and *compare* instructions, respectively. Recall that, in
the EDDI transformation, branches are the synchronization points. The redundant
instructions from 3 to 7 are introduced in order to compare the predicate p11 to its

Original Code	(a) EDDI+ECC+CF Code	(b) EDDI+ECC+ECF Code
add r11 = r12, r13	add r11 = r12, r13	add r11 = r12, r13
	1: add r21 = r22, r23	1: add r21 = r22, r23
cmp.lt.unc	cmp.lt.unc	cmp.lt.unc
p11, p0 = r11,r12	p11, p0 = r11,r12	p11, p0 = r11,r12
	2: cmp.lt.unc	2: cmp.lt.unc
	p21, p0=r21, r22	p21, p0=r21, r22
	3: mov r1 = 0	3: (p21) xor RTS=sig0, sig1
	4: (p11) xor r1=r1, 1	
	5: (p21) xor r1=r1, 1	
	6: cmp.neq.unc	
	p1,p0 = r1, 0	
	7: (p1) br faultDetected	(p11) br L1
(p11) br L1	(p11) br L1	. . . .
. . . .	. . . .	4: xor RTS=sig0, sig1
L1:	L1:	L1:
	8: xor GSR=GSR,L0_to_L1	5: xor GSR=GSR, RTS
	9: cmp.neq.unc	6: cmp.neq.unc
	p2,p0 = GSR, sig.1	p2,p0 = GSR, sig.1
	10: (p2) br faultDetected	7: (p2) br faultDetected
	11: cmp.neq.unc	8: cmp.neq.unc
	p3,p0=r11, r21	p3,p0=r11, r21
	12: cmp.neq.or	9: cmp.neq.unc
	p3,p0=r12, r22	p3,p0=r12, r22
	13: (p3) br faultDetected	10: (p3) br faultDetected
st m[r11] = r12	st m[r11] = r12	st m[r11] = r12

Fig. 2.18 (**a**) Control flow checking using software signatures [34] and (**b**) Enhanced control flow
checking [28]

duplicate p21 and branch to error code if a fault is detected. At instruction 8, the control flow additions start that transform the GSR from the previous block signature to the signature for the currently executing block. The instructions 9 and 10 confirm if the signature is correct, in case an incorrect signature is detected, the error code is invoked. Finally, instructions 11–13 are inserted to handle the synchronization point induced by the later store instruction. This transformation will detect faulty control flow transfers between two blocks which are not unauthorized. Any such control transfer will result in an incorrect signature no matter if the erroneous transfer jumps to the middle of a basic block. These issues have led to the enhanced control flow protection approach [28]; see example in Fig. 2.18b. In this technique, for all the blocks, a dynamic equivalent of a run-time adjusting signature is used (also for the basic blocks which are not control flow merges). Each block asserts its target while using the run-time adjusted signature, and in response each target confirms the transfer by checking the GSR. Figure 2.18b demonstrates how the enhanced control flow checking works. Similar to the previous control flow checking example, in this example instructions 1 and 2 are the redundant duplicates for the add and compare instructions, respectively. The run-time signature for the target of the branch is computed via Instruction 3 by *xor*'ing the signature of the current block with the signature of the target block. As the branch is predicated, the assignment to RTS (Run-Time Signature) is also predicated using the redundant register for the predicate. Instruction 4 is the equivalent of instruction 3 for the fall through control transfer. To compute the signature of the new block, instruction 5 at the target of a control transfer *xors* RTS with the GSR. Afterwards, at instruction 6, this signature is compared with the statically assigned signature, in case of mismatch an error code is invoked with instruction 7. Just as before, instructions 8 through 9 implement the synchronization checks for the store instruction.

The SWIFT approach [28] is demonstrated in Fig. 2.19 where all the instructions are duplicated and before the *store*, the comparison instructions for fault detection are placed. In this case, it is only the *store* instructions which ultimately send data out of the *SoR* (Sphere of Replication). The system will function correctly, as long as it is ensured that *store* instructions execute only if they are "meant to" and the *store* instructions write the correct data to the correct address. This observation is used to restrict enhanced control flow checking only to blocks having the *store* instructions. In this scenario, the updates to the GSR and RTS are performed in all

Fig. 2.19 Software implemented fault tolerance (SWIFT) [28]

Original Code	Transformed Code [SWIFT]
add r1 = r2, r3	add r1 = r2, r3
	1: add r1' = r2', r3'
mul r1 = r1, 8	mul r1 = r1, 8
	2: mul r1' = r1', 8
	3: br faultDet, r1 != r1'
	4: br faultDet, r2 != r2'
st [r1] = r2	st [r1] = r2

blocks; however, the comparisons for signatures are restricted to blocks with *store* instructions. With this optimization, if the signature check instructions are eradicated, this will further alleviate the overhead for fault tolerance with no reduction in the reliability. Since signature comparisons are computed at the beginning of every block that contains a *store* instruction, any deviation from the valid control flow path to that point will be detected before memory and output is corrupted. This optimization has relatively lesser negative impact on the performance compared to the EDDI. Both the branch checking and enhanced control flow checking are somewhat redundant. While branch checking makes sure that branches are taken in the proper direction, the enhanced control flow checking ensures that all control transfers are made to the proper address. However, *verifying all control flow includes the notion of branching in the right direction.* Therefore, performing the control flow checking alone is adequate to detect all control flow errors.

Besides fault detection, a reliable system requires fault recovery, too. The SWIFT transformation can be seen as a DMR-like implementation that provides fault detection, but not recovery. For recovery, a TMR-like implementation SWIFTR (*Software Implemented Fault Tolerance with Recovery*) approach is presented in [71] that employs triplicated instructions and majority voting. Such full-scale redundancy solutions, however, incur significant power, area, and performance overheads. To alleviate these overheads, there are some techniques that offer enhanced control flow protection like CRAFT [28, 71] and Instruction Vulnerability Factor-based techniques [76] via duplicating only the critical instructions, i.e., instructions that have a relatively high probability to lead to a software failure/crash in case of a soft error, for instance, *load*, *store*, *jump*, *branches*, and *calls*. However, these techniques incur additional >40 % performance loss, increased register pressure due to more register usage, and excessive memory overhead because of instruction and data redundancy [28]. Furthermore, an increased number of critical instruction executions may lead to excessive rollbacks during recovery because of an increased probability of software failures and fault propagation to/from memory, when a fault occurs in the hardware of the memory pipeline stage [28]. Besides offering protection only at the instruction level, some advanced work of [77] exploits the unused bits of a register for in-register duplication, while [80] performs both the instruction and data duplication. These redundancy-based techniques incur a significant performance/memory overhead ($>2x$–$3x$) [27, 28, 77, 80, 92], which is typically prohibitive for embedded systems.

Besides excessive performance overhead, one of the primary issues of instruction redundancy and scheduling techniques, like [27, 28, 72, 73], is that they treat all instructions in the same way. This is because their software-level reliability estimation models (Register Vulnerability Factor[1] [73] or Program Vulnerability Factor[2] [74, 75]) do not distinguish between different types of errors in the software caused by the hardware-level faults during the execution

---

[1] Register Vulnerability Factor considers the register live period as a measure for the reliability.

[2] Program Vulnerability Factor relates the software reliability to the bits for Architecturally Correct Execution in different programmer-visible architectural components (Register File, ALU, etc.), but hides the physical components (e.g., there are 256 physical registers, but 32 are visible to the programmer).

of different instructions that use diverse processor components in different pipe-line stages. Moreover, these models are computed without considering the processor architecture. As a result, software-level reliability techniques of this kind are not very efficient. For vulnerability reduction, different compile-time approaches have evolved that seem to have promising effects on lowering the error probability of the software programs. For example, the instruction scheduling phase during the compilation can impact the instruction vulnerability by affecting the vulnerable periods of instructions and their operands in different pipeline resources. Towards this, several compile-time reliability-aware instruction scheduling approaches have been proposed [27, 76] that reorder the instruction profile of a program while incurring relatively limited performance degradation and almost no memory overhead compared to instruction redundancy techniques. The work of [77] minimizes the residency cycles of vulnerable bits inside the issue queue of superscalar processor by performing instruction scheduling at run-time. However, this technique requires architecture modification of the hardware scheduler and introduces a significant hardware overhead. In contrast, ISSE [81] reschedules a program's assembly code to minimize the operands' vulnerable periods via exploiting the slack time. The works in [78, 79] perform instruction rescheduling after the performance-optimized scheduling in order to reduce the vulnerable periods of registers. The slacks are identified after a performance-driven instruction scheduling, which already tries to minimize the slacks as much as possible to avoid pipeline stalls to improve performance. As a result, state-of-the-art instruction scheduling techniques [27] and [76] provide limited reliability improvements of 2 % and 9 %, respectively. Furthermore, the error probability is reduced by lowering the vulnerability of register file [80] or software program [28] through minimizing the register lifetime. These state-of-the-art solutions offer limited reliability improvements as they primarily improve the reliability of the register file, which typically covers a small portion of the processor layout compared to the pipeline and instruction execution unit, thus ignoring the complete processor perspective.

2.6 Summary of Related Work

In this chapter, the background related to various reliability threats, i.e., soft error, NBTI-induced aging effect, and process variation is discussed. The mechanisms of these reliability threats, their sources, and how they are modeled are presented. Since the focus of this manuscript is on soft errors, a detailed literature survey regarding the soft error estimation and mitigation techniques is presented at various levels of system abstraction, i.e., circuit level, architecture level, and program level.

Although there has been plenty of hardware-level software mitigation works at the device, circuit, and architectural layers, these techniques are not area- and power-wise cost effective as they incur extra circuitry besides their high verification/validation costs [19, 33, 71, 107, 116]. To alleviate this overhead, various soft error mitigation techniques at the software level have evolved. The control flow checking and instruction/register value duplication result in significant performance and memory overhead, while register vulnerability reduction techniques provide limited reliability improvement. In contrast, instruction scheduling for reliability reorders the instructions of a software program without costing memory overhead and with limited/no performance overhead [73, 81]. However, these techniques ignore the complete processor perspective, as they only try to reduce the vulnerability of the register file that covers only a small portion of the processor layout compared to the complete pipeline. As a result, these techniques [73, 81] provide limited reliability improvement (2–9 %). State-of-the-art compiler-level reliability techniques have not exploited the prospective opportunities which exist at the compiler front-/middle-end optimizations that may impact the software code for improving reliability with reduced performance overhead. Furthermore, the slacks for reliability improvement are identified after a performance-driven instruction scheduling that already minimized slacks to avoid pipeline stalls to improve performance. As a result, state-of-the-art instruction scheduling techniques [73] and [81] provide limited reliability improvements of 2 % and 9 %, respectively.

In the following chapter, a comprehensive view of the novel contributions of this manuscript is presented along with details on the developed concepts, techniques, different design challenges, and motivating error analysis at the application software program level while considering the hardware-level faults.

Chapter 3
Cross-Layer Reliability Analysis, Modeling, and Optimization

The main problem targeted in this manuscript is to reduce the software programs' susceptibility to soft errors on unreliable or partially reliable hardware, and to improve the reliability of the overall system. This needs to account for the knowledge/information from both the hardware level (i.e., where the faults occur) and the software level (i.e., where the errors are observed). The problem gets even more challenging when considering optimization under tolerable performance overhead constraints, as typical for embedded computing systems. However, mitigating the hardware-level faults at the software level is not straightforward and poses several research challenges with respect to modeling and optimization at the respective software layers. This chapter presents a comprehensive overview of the proposed *cross-layer reliability analysis, modeling, and optimization flow* (Sect. 3.1) to address these research challenges. In particular, the goal is to enable reliable code generation and execution on unreliable hardware. The techniques at different system layers along with their interactions are highlighted.

In order to enable a cross-layer modeling and optimization flow, important information at different system layers needs to be identified and exchanged across layers in order to improve the accuracy of reliability modeling and efficiency of reliability optimization at the corresponding abstraction. State-of-the-art software reliability analysis and optimization techniques ignore the hardware-level knowledge (i.e., fault distribution, fault probabilities, and spatial/temporal effects for different instruction types) and do not analyze fault impact on instructions' execution in pipeline stage by stage, and thus lack potential for optimization at the corresponding levels of granularity. Therefore, a key challenge is to *bridge the gap between hardware and software* for quantifying hardware-level faults at the software level and enabling corresponding optimization potential. An associated challenge is to understand different reliability related factors and to identify different types of errors when faults are injected in different processor components during the execution of different types of instructions and different application software programs. This analysis would help in establishing a relationship between different instruction types and faults in different processor components. Towards this end, this chapter

S. Rehman et al., *Reliable Software for Unreliable Hardware*,
DOI 10.1007/978-3-319-25772-3_3

presents detailed error characterization and a comprehensive software program-level reliability analysis (Sect. 3.2), which is performed using fault injection experiments at different fault rates. The goal is to analyze the execution of different instructions in the presence of faults in different hardware components, and to understand the spatial/temporal aspects of reliability and the distribution of different types of program errors when instructions execute through different pipeline stages. Further, error analysis for different applications is performed to study the impact of different instruction profiles. The observations from this analysis help in identifying several important parameters which will be leveraged to devise cross-layer software reliability models (Sect. 3.3) considering the hardware-level knowledge and thereby bridging the gap between hardware and software.

In order to enable reliability optimization at different system layers, these reliability models need to quantify the software programs' reliability at different levels of granularity (i.e., instruction and function/task) that are adapted to the respective levels of the system design abstractions. For instance, instruction-level reliability quantification will enable reliability optimizations at the compiler level, while function/task-level reliability quantification will enable optimizations at the system software level. State-of-the-art has not fully explored the reliability optimization potential at the compiler and system software layers due to the unavailability of the instruction-level reliability models. Towards this end, a cross-layer reliability optimization flow is proposed that facilitates interactions between the reliability-driven compiler (Sect. 3.5) and the reliability-driven system software (Sect. 3.6) to provide reliable code generation and execution. For reliable code generation, the goal is to enhance the reliability of application software programs through (1) reducing error probabilities by lowering the instructions' vulnerabilities to soft errors; and (2) selective error detection and recovery enabled by instruction-level redundancy under constrained scenarios. Multiple reliable code versions for a given application program can be generated that provide diverse reliability and performance properties. This can be leveraged by the reliability-driven system software for reliable application execution. Further reliability improvements can be obtained by exploiting the architecture-level support for redundant multithreading. The scenario becomes more challenging when multiple concurrently executing applications compete for resources in a many-core system to provide redundant threading support. In this context, a key challenge is to jointly account for redundant multithreading, multiple reliable versions, and resilience properties of different applications. To address this, a novel reliability-driven run-time system is presented towards the end of this chapter.

Note, this chapter provides a comprehensive overview of the proposed contributions, their interactions in a cross-layer flow, discussion on the associated challenges, and enabling program reliability analysis. The detailed techniques, algorithms, and an in-depth reliability analysis of the proposed cross-layer modeling and optimization techniques will be discussed in Chaps. 4 and 5, respectively. Furthermore, in Chap. 5, individual reliability optimization techniques will be evaluated by comparing these with state-of-the-art, while the overall cross-layer reliability optimization methodology will be compared to different single-layer solutions in Chap. 7.

3.1 Cross-Layer Reliability: System Overview

Considering the increasing reliability threats with technology scaling and the growing mitigation costs, cross-layer reliability optimization is important to achieve cost-effective reliability such that different system layers contribute their best towards it [19, 107]. Furthermore, different system layers exhibit inherent error masking capabilities that can be leveraged in a cross-layer optimization flow to achieve cost-effective reliability. Cross-layer approaches engage two or more, adjacent or nonadjacent, hardware and/or software layers to mitigate reliability threats, such that different layers exchange information or adapt each other at design-/run-time to achieve high efficiency in terms of resources, performance, power/energy, and/or cost. Different layers may contain faults and avoid propagation to the upper system layers by exploiting their inherent error masking capabilities and reliability mitigation techniques amenable to these layers, such that these errors do not get visible to the application/device user. The key challenge is to identify those error masking potentials, model reliability at appropriate granularity and system abstraction level, and to devise efficient reliability optimization techniques while leveraging the information exchange across multiple layers. Furthermore, to enable cross-layer reliability optimizing techniques, there is a need for cross-layer software reliability modeling and program-level error manifestation analysis to understand how these hardware-level faults manifest at the software program level. In the following, a novel cross-layer reliability analysis, modeling, and optimization flow is proposed and discussed in detail in the later subsections of this chapter.

Figure 3.1 presents a high-level overview of the proposed cross-layer reliability modeling and optimization system and Fig. 3.2 presents the interactions between the components of different layers. The novel contributions of this manuscript are highlighted, i.e., cross-layer reliability model and estimation, dependable code generation through reliability-driven compiler, and dependable application execution through reliability-driven offline and run-time system software. In the proposed cross-layer reliability modeling and optimization flow, different layers interact with each other and exchange information for reliable code generation and execution on an unreliable or partially reliable hardware. The goal is to leverage multiple system layers in an integrated fashion for reliability optimization under user-provided tolerable performance overhead constraints. Enabling efficient reliability optimizing techniques also require appropriate models that exploit the knowledge from both software and hardware layers to efficiently model the reliability factors at the corresponding system layers adapted to the appropriate granularity, for instance, instruction and function/task level.

In the following, a brief overview of the overall flow is presented with more details in the subsequent section.

1. **Cross-Layer Software Program Reliability Modeling (Chap. 4)**: In order to enable efficient cross-layer software reliability optimization, there is a need for accurate reliability modeling at the appropriate granularity. In a cross-layer software reliability modeling flow, there is a need to bridge the gap between the

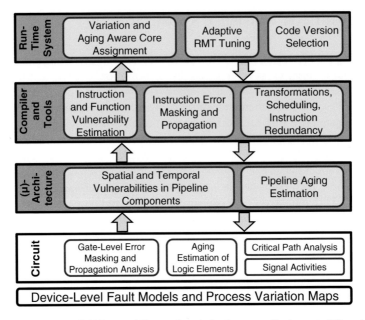

Fig. 3.1 Cross-layer reliability modeling and optimization: contributions at different system layers

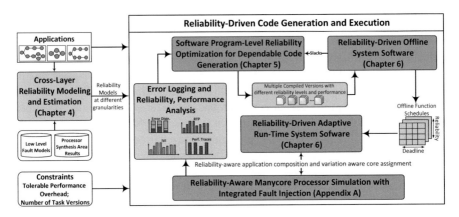

Fig. 3.2 System overview for reliable code generation and execution engaging different system layers

hardware and software by quantifying the effects of hardware-level faults at the software level, while accounting for the knowledge of the processor architecture and layout. This manuscript performs a cross-layer reliability analysis and leverages this to devise novel cross-layer software reliability modeling and estimation techniques. The analysis helps in understanding how hardware-level faults manifest at the software layer, especially considering their relationship to the fault location (i.e., in which processor components a fault occurs) and the instruction type (i.e., during the execution of which instruction a fault occurs); see Sect. 3.2 for detailed reliability analysis and error characterization. This analysis and characterization will be exploited to identify different reliability-affecting hardware- and software-level parameters and to formulate different software-level reliability models for quantifying error probabilities, error masking, and propagation effects.

2. **Software Program-Level Reliability Optimization for Dependable Code Generation (Chap. 5)**: In the proposed cross-layer reliability optimization flow, first, the application software code is compiled using a traditional compiler flow with performance optimizations. Afterwards, the application binaries are executed on a reliability-featured manycore processor simulator, where the application program is analyzed for reliability using the above-discussed reliability metrics, different error types are obtained after extensive fault injection experiments, and performance properties are analyzed. The *reliability-driven code generation* optimizes the application code for reliability using different reliability-driven transformations, instruction scheduling, and selective instruction redundancy techniques. These reliability-driven code-generation techniques may incur additional performance overhead due to, for instance, redundant instructions. Therefore, these reliability-driven transformations and instruction redundancy are employed under tolerable performance constraints that can either be provided by the user or obtained through slack analysis done at the system software layer for different functions. The tolerable performance overhead is distributed among different application functions depending upon their resilience/vulnerability properties. The reliability-driven transformations and instruction scheduling aim at reducing the error probability while the selective instruction redundancy technique targets error detection and recovery. A set of different compiled versions for each application function is generated, such that different functions differ in terms of their reliability and performance properties. Optionally, a multi-pass optimization loop can be employed, where these reliable function versions can be forwarded again to the offline system software for iterative improvement and slack analysis to prune/optimize the reliability-performance design space of multiple function versions. These reliable function versions are then forwarded to the reliability-driven offline and run-time system software layers for dependable application execution.

3. **Reliability-Driven System Software (Chap. 6)**: First, the *reliability-driven offline system software* performs reliability-driven application composition and function scheduling. It generates multiple offline function schedules, where each function schedule represents a particular dependable application composition using a different set of reliable function versions. It estimates the *Reliability-Timing Penalty* of different function versions and selects the one that minimizes the overall *Reliability-Timing Penalty* considering the reliability and execution time of the previously executed functions and the remaining time until the application deadline. These function schedules are then forwarded to the *reliability-driven run-time system software* that manages dependable application execution on single core or multi-/manycore processors considering core-to-core frequency

variations. Towards this end, the proposed cross-layer reliability optimization flow employs adaptive *Dependability Tuning* at the run-time system software layer. The *Dependability Tuning* enables dependable application execution through resilience-driven adaptive control of redundant multithreading, variation-aware thread-to-core mapping, and reliable function version selection.

Summary: In order to enable cross-layer reliability solutions (with a focus on software programs), reliability modeling and optimization techniques are developed across different layers. To devise efficient reliability models and optimization techniques, in the following, detailed program error analysis is presented. This analysis is important to understand how the hardware-level faults are manifested at the software level and helps in identifying the vulnerability properties of different instructions (or in other words, vulnerable regions inside the program code) and their relationship to different types of errors when faults occur in different processor components. This analysis helps in exposing various important parameters that contribute towards developing cross-layer reliability models.

3.2 Software Program-Level Reliability Analysis

In order to analyze the effects of hardware-level faults at the application software program level, it is important to understand how these hardware-level faults occur and how their effects propagate to the program level and how they manifest in the form of different types of errors. To illustrate this, various fault injection experiments were performed (see Appendix A for details on the experimental setup), such that bit flip faults are injected in different hardware components during the execution of different instructions and the resulting errors in the program's output are monitored. Figure 3.3 illustrates two example fault injection experiments for the ADPCM application executing on an embedded LEON2 processor (*SPARC v8* architecture [35]). The faults are injected in the register file and the pipeline as shown in Fig. 3.3.

Fault Injection Experiment 1: A fault is injected in the register file during the execution of *"instruction 0x18c."* In case the particle strikes the register *g2* during the

Fig. 3.3 Illustrating the impact of faults in different processor components and their impact on the program execution (layout of LEON2 from [35])

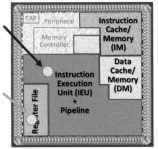

[Photo: Gaisler @ IEEE DSN'02]

execution of this instruction, a corrupted register value did not affect the correctness of this final output of a program. This hints towards the inherent *error masking* capabilities of an application software program. In case of other applications, where such an error in the register file propagates to the final output of a program, the error can be *tolerable* or *non-tolerable*. For instance, in case of image/video processing applications this may be tolerable while for security applications like AES this may not be tolerable.

Fault Injection Experiment 2: In this experiment, the fault is injected in the Pipeline + Instruction Execution Unit part of the processor as highlighted in Fig. 3.3. In this case, the particle strike causes a bit flip in the instruction decoder that can have one of the following potential effects: (1) The *add* instruction may be interpreted as any other instruction, e.g., *load* or *store* instruction due to the corruption of the opcode in the instruction word. (2) The *add* instruction may be interpreted as a *non-decodable* instruction due to the corruption of the opcode in the instruction word. (3) Different operands for the *add* instruction may be selected in case the operand bits of the instruction word are corrupted. Such errors may lead to Application Failures (like crash, hang, and abort; see error characterization in Sect. 3.2.1), for instance, in case (1) the *load/store* instructions may lead to memory access errors; in case (2) the *non-decodable* instruction may lead to processor stall or application hang due to an unrecognizable instruction type; and in case (3) if the *add* instruction corresponds to an address computing instruction, then its corresponding *load/store* instructions may experience a memory access error (e.g., out of bound memory access) due to accessing a potentially incorrect address.

The above-discussed experiments illustrate that a corrupted register value for an output data computing instruction may have a less severe impact compared to a corrupted opcode or corrupted address value. Which part of the program is affected by soft errors and what type of error manifests at the program layer depend upon the following factors: (1) *the fault location at the hardware layer*, i.e., in which processor component a fault has happened; (2) *the fault distribution*, i.e., the probability of fault in different processor components; and (3) *the instruction type*, i.e., during the execution of which instruction a fault has happened. This in turn depends upon the instruction profile of different applications. Therefore, different applications exhibit varying susceptibility to soft errors even under the same processor hardware and same fault scenario.

In Sect. 3.2.2, detailed error distribution analysis is performed to study the impact of hardware-level faults on the software program level in detail. This analysis will provide several observations for investigating different software-level reliability models and for categorizing instructions with respect to their importance to different error types. Towards this, different types of errors are defined below.

3.2.1 Error Characterization

Following the well-established error characterization in the literature [91], the following error types are adopted in this manuscript that are then further classified into more detailed error subcategories.

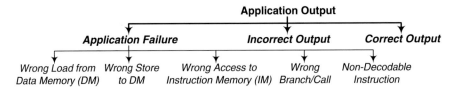

Fig. 3.4 Different types of manifested errors

After the execution under a given fault scenario, the output of an application can be classified into one of the following three main categories (see Fig. 3.4).

1. **Correct Output**: The application output matches 100 % with that of a "golden run," i.e., a fault-free application execution.
2. **Incorrect Output**: The application output does not match 100 % with the output of a "golden run." The errors in the Incorrect Output could be due to the following reasons: (a) *Silent Data Corruptions* that correspond to errors which are not detected even in the presence of a certain error detection mechanism and corrupt the final program output; and (b) *Detected Unrecoverable Errors* that correspond to errors which are detected but not recovered due to, for instance, unavailability of recovery mechanisms. For the ease of discussion, all errors that propagate to the final output of an application software program are treated as Incorrect Output errors.
3. **Application Failure**: The application software program did not finish its execution successfully due to a potential abort, crash, hang, or exception. This corresponds to a fatal application behavior, which is in particular not acceptable in critical embedded systems. The Application Failure is further categorized in the following different error subcategories.

   (a) *Wrong Load from Data Memory (DM)*: A segmentation fault occurs because a corrupted load address is accessed from the data memory.
   (b) *Wrong Store to Data Memory*: A segmentation fault occurs because a corrupted store address is accessed in the data memory.
   (c) *Wrong Access to Instruction Memory (IM)*: This is due to a segmentation fault which occurs because of a faulty Program Counter (PC) or Next Program Counter (NPC).
   (d) *Wrong Branch/Call*: The result of the jump/branch/call target calculation is erroneous and therefore a jump/branch/call is made to an incorrect address that halts or aborts the application execution.
   (e) *Non-Decodable Instruction*: A fault in an opcode field of an instruction word may lead to a non-existing opcode that leads to a potential abort of the application execution.

Note, this error categorization follows the application user perception, which is relatively important from the software perspective, i.e., the user gets correct output, or the user gets Incorrect Output that may or may not be acceptable depending upon the application type (e.g., multimedia or security), or the user experiences an application execution failure that can either be an unintended termination of the application or a non-terminating execution run. From the software perspective, Application Failures basically mean that the application software fails to continue further execution due to different reasons, e.g., a crash, hang, and abort, and the result of the application execution is always unacceptable, i.e., describing the fatal application behavior. Both Application Failures and Incorrect Outputs are important and need to be detected/corrected for a highly reliable system. Considering the availability and quality of service issues, in several cases, avoiding Application Failures is one of the most desirable features to improve the system dependability and these failures should be avoided/handled first to keep the application execution alive [28, 92]. Moreover, recovery from Application Failures is typically very time consuming (i.e., high Mean-Time-To-Recovery) and affects the users' desired responsiveness and may even violate the performance constraints/deadlines [120]. Such cases may even be more critical in safety related applications (e.g., delay in the antilock braking system in automotives due to application restarting from a failure state). In such scenarios, it is highly inefficient to face several Application Failures. Similar explorations and observations can be found in state-of-the-art works like [28, 91, 92]. Furthermore, an erroneous output of an instruction can propagate to the inputs of multiple instructions and subsequently to multiple outputs of the software program. This in turn leads to more or less Incorrect Outputs with higher or lower error magnitudes or Application Failures in case of, for instance, address corruption.

In the following, a detailed error distribution analysis for different applications under three different fault rates is presented. The details on the fault injection parameters and fault injection procedure can be found in Appendix A.

3.2.2   Error Distribution Analysis

This section presents a detailed error distribution analysis for different applications from MiBench [111]. The analysis is performed using a reliability-driven processor simulator with an integrated fault injection module. To realize scenarios under various operational conditions, several parameters are taken into consideration leading to different fault rates (see parameter details in Appendix A). In this analysis, three fault rates are used: 1, 5, and 10 f/MCycles considering an embedded processor with *SPARC v8 instruction set* architecture operating at a frequency of 100 MHz. Figure 3.5 illustrates the distribution of different errors for various applications executing on a LEON2 embedded processor subjected to different fault rates. Figure 3.6 shows the instruction execution profile and detailed distribution of different types of Application Failures and Incorrect Outputs for some example applications.

The summary of key observations from the above analysis of Figs. 3.5 and 3.6 is given below.

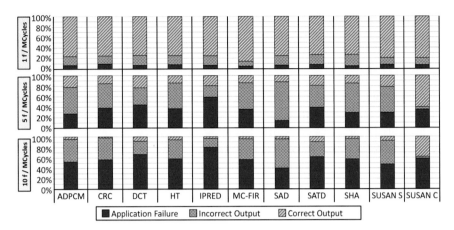

Fig. 3.5 Analyzing the error distribution at different fault rates

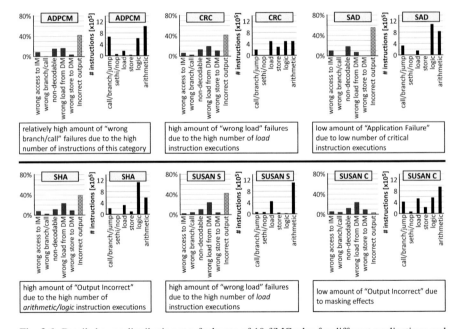

Fig. 3.6 Detailed error distribution at a fault rate of 10 f/MCycles for different applications and the corresponding instruction distribution

1. Different applications exhibit varying computational properties and instruction execution profiles and thereby exhibit dissimilar distribution of error categories. For example, the Application Failure due to a wrong load from the data memory or due to a wrong store to the data memory increases in a distinct way for the growing fault rate. In *ADPCM* and *SAD*, when going from 1 to 5 f/MCycles, the Incorrect Output cases grow. This is because of a large number of *arithmetic/logic* instructions. There are

fewer failures due to the reduced number of *load* and *store* instructions. However, when going from 5 to 10 f/MCycles, the share of Incorrect Output reduces because the amount of Application Failures increases due to instruction decoder faults, i.e., non-decodable instructions. Moreover, a large number of Application Failures are caused by corrupted loop exit conditions due to more control flow instruction executions.

2. For *SHA* and *SusanC*, the "wrong load" category is fairly prominent because of more load instruction executions, while the failures due to "wrong store" vary.

3. The failures for "wrong load from the data memory" happen primarily due to: (a) the bit flips in the operand containing the address during the memory pipeline stage, or (b) the bit flips in the Address Generation Unit during the address computation in the execute pipeline stage.

4. Since all instructions use instruction fetch unit and instruction decoder, the failures during the instruction-fetch (i.e., wrong access to instruction memory) and instruction-decode stages occur with similar probability for all instructions. For instance, if a bit flips in the opcode field of an instruction word, this may lead to a non-decodable instruction.

5. The key difference in the usages of the processor components comes from other pipeline stages, like execute and memory. Bit flips in the operands containing the address of a branch/call, load/store from/to a corrupted location (e.g., due to address corruption) of data memory, and wrong access to the instruction memory are typically not tolerable as these events may result in an Application Failure. In contrast, bit flips in the operands of arithmetic instructions (except address generation) may lead to an Incorrect Output error that may or may not be tolerable depending upon the resilience of a functional block [108, 109]. For instance, it may be tolerable in some functional blocks of a multimedia application (like pixel difference computation in SAD or frame/picture buffer [108–110]), but may not be tolerable in sensitive functional blocks like bitstream processing blocks.

Summarizing the above analysis and discussion, it is important to identify the reliability-wise importance of different instructions, which may vary in two different aspects: (1) instruction classification w.r.t. the type of error due to a fault during its execution; and (2) quantification of error probabilities for different instructions w.r.t. the area of processor components used by these instructions and the amount of time spent in these components. In the following, these two key reliability aspects of instructions are discussed in detail.

3.2.3 Instruction Classification: Critical and Noncritical Instructions

Depending upon the type/severity of an error (i.e., Application Failure or Incorrect Output) as a result of faults in the pipeline stages (other than instruction-fetch and instruction-decode) during its execution, an instruction is categorized as a "critical instruction" or "noncritical instruction."

Noncritical Instructions: If a fault during an instruction's execution leads to an Incorrect Output value but does not terminate or halt the software program execution, it is classified as a *noncritical instruction*. Examples of such *noncritical instruction* are arithmetic and logical data computing instructions except address generation ALU instructions (for instance, *add*, *sub*, *and*, *or*, etc.).

Critical Instructions: In contrast, an instruction is categorized as a *critical instruction* if a fault during its execution leads to an Application Failure, which depends upon the application control and data flow, instruction dependencies, and the associated error types. Potential examples of such *critical instructions* are control flow instructions (for instance, *jump*, *branches*, *calls*, etc.), memory-related instructions (like *load* and *store*), and all of their associated predecessor instructions (like arithmetic and logical instructions used for address generation) [112]. For a given fault rate, a relatively large number of *critical instruction* executions may lead to a high susceptibility towards Application Failures. In general, even for multimedia applications, faults in the execution of *critical instructions* cannot be tolerated. Similar observations are also made by several other works like [28, 91, 92]. Therefore, the protection of *critical instructions* is relatively more important.

Besides the knowledge of the *critical* and *noncritical* instructions, it is important to understand which instruction leads to which type of error in the application software. As discussed above, the error type is dependent upon the processor component in which the fault occurs and the instruction type that uses the pipeline resources in distinct ways for a different amount of cycles. Towards this end, the next section will introduce the notion of **spatial vulnerability** and **temporal vulnerability** as the area-wise and time-wise error probabilities, respectively, while relating to the area of processor components and the amount of time spent by different instructions in these components.

3.2.4   Spatial and Temporal Vulnerability

In order to quantify the application software program's reliability considering the underlying hardware, the following two parameters need careful investigation.

Spatial Vulnerability: Different processor components (i.e., register file, instruction and data memories, instruction decoder, execution units) cover different area of the chip footprint. Considering the case of soft errors where the number of particle hits is proportional to the surface area, the probability of a fault in a certain processor component is a function of its area, i.e., a processor component with a bigger area is more susceptible to particle strikes. Figure 3.3 illustrates that the probability of a particle strike on the Pipeline + Instruction Execution Unit is higher than that on the register file. This denotes the area-wise vulnerability of an instruction towards soft errors. Therefore, the *Spatial Vulnerability* of an instruction is defined as the probability of the occurrence of a fault during the execution of an instruction depending upon the area of the specific processor components that it uses.

Temporal Vulnerability: Different instructions spend varying amount of time (in terms of cycles) in different pipeline stages due to their distinct execution latency, instruction dependencies, different number of operands, and potential pipeline stalls. This manifests as varying vulnerable time periods in different pipeline stages. Therefore, the *Temporal Vulnerability* is defined as the probability of occurrence of a fault during the execution of an instruction depending upon the vulnerable time periods it spends in different processor components considering the above factors.

An Example: Figure 3.7 demonstrates the above-discussed concepts of spatial and temporal vulnerabilities with the help of three instructions *add*, *multiply*, and *load* and their execution in different pipeline stages considering a 5-stage integer unit

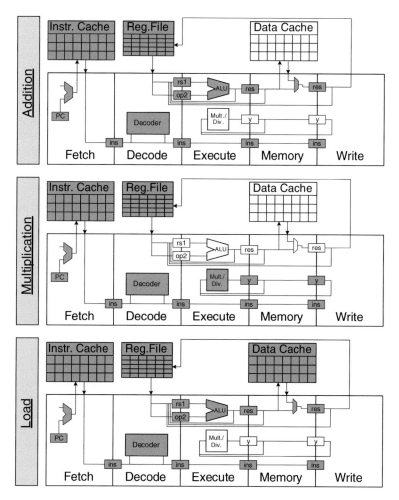

Fig. 3.7 Spatial and temporal vulnerabilities: different instructions using diverse processor components in pipeline stages

pipeline of the LEON2 processor. The components used by different instructions in different pipeline stages are denoted as filled boxes. Note that unlike the *load* instruction, the *add* and *multiply* instructions are not vulnerable in the memory stage. The vulnerability of *load/store* instructions in the execute stage is primarily due to the address calculation. The spatial vulnerability of the *load* instruction is higher compared to that of the *add* instruction due to the usage of more processor components.

Furthermore, the spatial vulnerability of the *multiply* instruction is higher compared to that of the *add* instruction due to bigger area of the multiplier unit compared to that of the Arithmetic and Logical Unit (ALU). However, simply considering the area of different processor components is not sufficient for accurate spatial vulnerability quantification because the probability that a particle strike manifesting as a soft error also depends upon the circuit characteristics of the processor component. For instance, a fault at a particular gate may be masked by the subsequent gates in the circuit; this is called circuit-level logical masking. A component with a high probability of fault denotes that a particle strike will most likely manifest as a bit flip. Moreover, a fault may be latched in a register but in a combinational circuit, it may be overwritten in the subsequent clock cycle (see further details on the gate-level soft error masking analysis details in Chap. 4). *In short, the quantification of error probability as a function of spatial vulnerability in different processor components (i.e., underlying hardware) depends upon their area and circuit-level error masking characteristics.* The temporal vulnerability of the *multiply* instructions is also more than that of the *add* instruction because the *multiply* instruction requires three cycles or more for execution, while the execution latency of an *add* instruction is one cycle.

Further Considerations for Spatial and Temporal Vulnerabilities from the Application Software Program's Perspective: When considering the complete instruction profile of an application, varying execution latencies have an impact on the preceding instructions in the pipeline as multi-cycle instructions stall the pipeline, thus increasing the temporal vulnerability of preceding instructions. Furthermore, the longer execution time of instructions also results in longer intervals between the variable usages (stored in the register file), which results in an increased temporal vulnerability w.r.t. the register file. From the application's perspective, increased spatial vulnerability can also be accounted by the usage of more variables stored in more live registers. Furthermore, it is important to consider that not all bits of operand variables are vulnerable for ensuring the correct software execution due to architecture-level or software-level error masking. That is, a fault in the hardware does not necessarily lead to an erroneous output of the application software program (or at least not in the user-visible error range) depending upon the instruction type, operand values, and control flow properties. A bit of an operand is characterized as a *vulnerable bit* if a bit flip in this bit leads to an error in the program's output; otherwise, it is characterized as a *non-vulnerable bit*. To demonstrate this concept of *vulnerable bits*, let us consider the following example.

```
RO = R1 & R2;    R1 = 32'bit x;    R2 = 0x0000FFFF;
```

In the above example, it is assumed that a fault may occur in *R1* or *R2*, but not in both. Note that a fault in the upper 16 bits of *R1* will not affect the value of *R0*. However, a fault in *R2* will affect the value of *R0*. Therefore, bits 0–15 of *R1* are *vulnerable bits*, while bits 16–31 of *R1* are *non-vulnerable bits*. In contrast to this, all 32 bits of *R2* are *vulnerable bits*. Therefore, besides the above-discussed parameters in spatial and temporal vulnerabilities, the knowledge from the *vulnerable bit* analysis is also important for devising the software program-level reliability models.

In summary, the reliability of a software program is a complex function of spatial and temporal vulnerabilities of its instructions and different error types that depends upon the instruction profile and application execution behavior. For instance, long vulnerable periods of register variables result in an increased *temporal vulnerability*, while using more live variables results in an increased *spatial vulnerability*. Moreover, even for the same overall performance, an increased *temporal vulnerability* of a *critical instruction* leads to a higher susceptibility towards Application Failure*s* compared to that of a *noncritical instruction*. The above-discussion and analysis provide hints and motivation towards consideration of different hardware and software parameters (like area and fault probabilities for different processor components, spatial and temporal vulnerabilities, and the knowledge about the *critical* and *noncritical instructions*) for software-level reliability modeling and optimization in constrained scenarios. In particular, since the spatial and temporal vulnerabilities are affected by the instruction profile, the reliability-driven code-generation techniques exhibit a potential towards software program's vulnerability reduction, as exploited in this manuscript (see Chap. 5).

3.2.5 Summary of Software Program Reliability Analysis and Relevant Parameters

The analysis in Figs. 3.5 and 3.6 illustrated that the Application Failures (for instance, branch errors due to wrong branch/call, memory access errors, non-decodable errors due to opcode bit flips, etc.) are critical program errors. Therefore, these errors need to be avoided (or at least minimized) as they may potentially terminate or block the application execution that may not be acceptable in various embedded systems deployed in critical application domains. The memory access errors can also be due to an erroneous execution of the predecessor critical ALU instructions for address generation that may lead to a wrong load/store to the data memory or a wrong access to the instruction memory. Therefore, reducing control flow and memory related *critical instructions* also leads to reduction in their dependent critical arithmetic and logical instructions used for address generation. A fault in the operand fields of the instruction word or the contents of the source and destination registers of branch, call, load, or store operations may lead to a potential abort/exception due to, for instance, out of bound memory access. Moreover, failures due to "INFINITE Loop" could be due to corruption of loop conditions and loop jump instruction. The Incorrect Output errors are mostly due to bit flips (in ALU, multiplier, etc.) occurring during the execution of arithmetic and logical instructions in the execute pipeline stage. As discussed above, the output error types vary depending upon the instruction type and the pipeline stage in which they occur. Therefore, in order to quantify the reliability at the instruction level, *spatial vulnerability* and *temporal vulnerability* of different instructions need to be considered. For a given fault rate, the probability of Application Failure*s* is directly proportional to the number of *critical instruction* executions and their *spatial and temporal vulnerabilities*.

In short, the above analysis is leveraged to derive different parameters for characterizing the reliability-wise importance of different instructions. The key parameters from the hardware and software perspective are: (1) the knowledge of **critical** and **noncritical** instructions, and (2) **spatial** and **temporal vulnerabilities** considering area of different processor components used by different instructions, probability of fault for different processor components depending upon their circuit design, and program-level knowledge of vulnerable bits and vulnerable periods that depends upon the instruction profile and schedule. The above-discussed parameters serve as an input for developing program-level reliability models at different granularity (i.e., instruction and function/task), which are adapted to the granularity of applying an appropriate reliability optimization technique. Furthermore, the software-level reliability techniques need to jointly account for the above-discussed hardware- and software-level information/parameters. Note that our fault injection experiments and fault modeling at the Instruction Set Simulator level also account for the above-discussed notion of spatial and temporal vulnerabilities and the respective values of the parameters.

3.3 Cross-Layer Reliability Modeling: A Soft Error Perspective

In the previous section, an extensive reliability analysis is performed to understand which instructions lead to which type of errors in the software program considering faults in different processor components and what their error masking behavior is. This analysis is leveraged to develop software program-level reliability models that account for both the hardware- and software-level knowledge. Towards this end, this manuscript addresses the key challenge of bridging the gap between hardware and software through a cross-layer reliability modeling flow that enables quantifying the effects of hardware-level faults at the software level, while accounting for the knowledge of the processor architecture and layout.

Figure 3.8 presents the proposed cross-layer reliability modeling flow with the information exchange across different layers for accurate estimation of reliability at the software level. The information from different layers of the system design abstraction (i.e., circuit, microarchitecture, and application) are used to devise

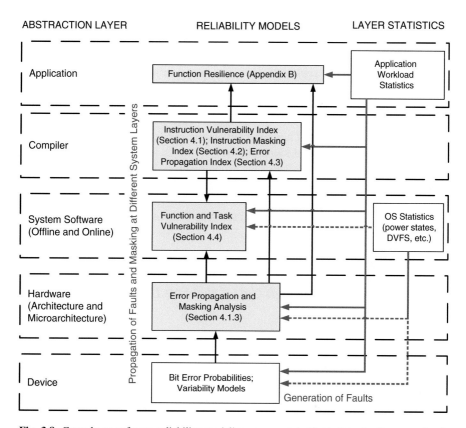

Fig. 3.8 Cross-layer software reliability modeling: an example illustration for the case of soft errors

important parameters required for developing accurate software program reliability models. These reliability models are then used to develop software program-level reliability optimization techniques. Therefore, these software program-level reliability models need to be adapted to different granularity (e.g., instruction and function/task levels) so that they can be applied at different system layers. The reliability threats such as soft errors and aging occur at the lower layer (i.e., at the device level) and their effects propagate upward through various system layers and finally appear at the application layer. Therefore, this manuscript adopts a *bottom-up approach* for reliability modeling and estimation. To bridge the gap between hardware and software, the proposed cross-layer reliability modeling flow leverages various parameters from both hardware and software layers. The identification of some important hardware-/software-level parameters has already been discussed above in Sect. 3.2. In the following, different parameters for reliability models are discussed, followed by an overview on the proposed reliability models at different granularity.

At the **hardware level**, the following important parameters are considered: (1) *Gate-Level Error Masking Probabilities* that are used to obtain the probability of fault (P_{fault}) for different processor components; and (2) *Spatial and Temporal Vulnerabilities*. At the **software level**, the following important parameters are considered: (1) *Basic Block Execution Probabilities* that are used to determine the error masking behavior at the application software program level. (2) *Instruction Type* which denotes the average-case error masking probabilities of different instructions, e.g., a bit-wise *and / or* instructions will mask 50 % of errors considering the input probability = 0.5 for "1" and "0." (3) *Register Live-In and Live-Out Information* is used for static estimation of temporal vulnerable periods for variables used in the static estimation of instruction vulnerabilities.

In order to provide an overview of the proposed cross-layer modeling flow, in the following, we present a brief description of different modeling efforts performed and parameters considered in this manuscript. Detailed techniques, methods, and results for these different models are discussed in Chap. 4.

3.3.1 Estimating the Probability of Fault for Different Processor Components Through Gate-Level Soft Error Masking Analysis

Probability of fault for different processor components is obtained through a detailed gate-level soft error masking analysis. The inputs to this analysis are circuit netlist, particle strike-to-error probabilities for logic and memory elements in the netlist, and signal probabilities. For a logic gate or memory element, particle strike to soft error probability is typically denoted as logic element error probability or bit error probability, respectively, as shown in the circuit layer in Fig. 3.8. These error probabilities are typically vendor specific and depend upon various technology parameters [113] and are typically unavailable under an open-access program. Finding these probabilities typically requires detailed SPICE-level simulations and can be done using the Predictive Technology Model [114] data. Since this is beyond the scope of this manuscript, without the loss of generality, in this manuscript these probabilities are considered as *1* due to the unavailability of this technology data. The signal probabilities can be obtained through gate-level simulations using, for instance, *ModelSim* [124] software.

It is important to note that not all faults in a logic element (i.e., gates) of a combinatorial circuit affect its output because a fault may be masked due to the logical masking effects of the subsequent gates. $P_{\text{fault}}(c)$ for a processor component c can be estimated as the average-case fault probability by considering an error at different gates in the circuit netlist of the component c and estimating the probability of an error at the circuit's output. For this, a conditional probability analysis of error propagation through multiple gates for a given circuit netlist is performed considering the signal probabilities at each gate which vary depending upon the input data. An alternate way is to perform the analysis for average-case signal probabilities. First, the masking probabilities of individual gates in the netlist are estimated. Afterwards, the error masking probability at a gate along a path (i.e., a set of connected gates leading to an output) is estimated. It is given as the probability that an error at this gate will not affect the correct output of this path, and it is calculated based upon the gate input signal probabilities and error probabilities. After finding the masking probabilities for all the paths, the final masking probability at the circuit output is obtained as the product of each of the paths' masking probability values. An exhaustive analysis of the netlist is typically very time consuming. The work in [115] proposes an approximation-based technique that trades off the analysis time with analysis accuracy.

Note, $P_{fault}(c)$ basically serves as an input to our software-level reliability models discussed below. Due to the unavailability of an open-source gate-level soft error analysis tool, a simple approach is developed in the scope of this manuscript to obtain this data. Other approaches like [50–52] can also be deployed to obtain this fault probability information as this is orthogonal to our software-level reliability models and our cross-layer reliability modeling flow provides a clean interface for incorporating this information in the form of a parameter $P_{fault}(c)$ for a given processor component c.

3.3.2   Estimating Reliability at the Instruction Granularity as a Function of Vulnerability, Error Masking, and Propagation

In order to quantify the reliability of a software program, three important characteristics need to be captured: (1) What is the probability that an instruction execution gets erroneous? (2) What is the probability that an erroneous instruction's output ultimately propagates to the final output of an application software program? (3) If the error is not masked, to how many final outputs does this error propagate to? In order to estimate the software program reliability considering the above three characteristics at the granularity of an instruction, i.e., the granularity which is important to develop instruction-level reliability optimizing techniques as adopted in compiler or application levels, this manuscript introduces three novel metrics, namely (1) *Instruction Vulnerability Index*, (2) *Instruction Error Masking Index*, and (3) *Instruction Error Propagation Index*. Figure 3.9 presents an abstract illustration of these three reliability metrics. In the following, we briefly describe the fundamental

Fig. 3.9 An abstract example illustrating the concept of error probability, error masking, and propagation at the instruction granularity

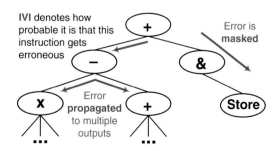

concept of these three instruction-level reliability metrics; further details on concept, methods, and results are discussed in Chap. 4.

IVI—Instruction Vulnerability Index: This model quantifies the error probability at the instruction level by jointly accounting for the spatial vulnerability (i.e., area-dependent error probability) and temporal vulnerability (i.e., time-dependent error probability) and the above-discussed probability of fault in different processor components. Quantifying these vulnerabilities for a given application software program requires the following two analyses: (1) *vulnerable bit analysis* for different pipeline components considering the fault probabilities and area of different processor components (like register file, ALU, and cache controller); and (2) *vulnerable time period analysis* because different instructions spend different time in processor components.

IMI—Instruction Error Masking Index: The output errors for several instructions may be masked due to data flow properties (depending upon the instruction type and the value of the operand variables) and changing control flow properties. To quantify these effects, an *Instruction Error Masking Index* model is developed that estimates the probability of an error at an instruction being masked until the visible output of an application software program.

EPI—Instruction Error Propagation Index: In case the error is not masked at an instruction, it can propagate to multiple outputs of the software program. The EPI model quantifies these error propagation effects for an error at an instruction using the non-masking error probabilities of the successor instructions. EPI provides a statistical estimate in the form of an accumulating function of non-masking probabilities.

The three above-discussed models jointly provide a measure of severity of an error at the instruction granularity. For instance, an instruction with a high IVI is not necessarily the reliability-wise most important instruction because it could also have a very high IMI and very low EPI value and most of the errors for this instruction may be masked due to the control and data flow properties. Therefore, the above-discussed models need to be jointly leveraged to determine the reliability-wise importance of different instructions in a given data flow graph for an application software program. It is important to note that in order to facilitate compiler-level reliability optimizations (see Chap. 5), IVI, IMI, and EPI are estimated statically using the control and data flow graph (CDFG) of an application software program. The static analysis accounts for basic block execution probabilities, average-case masking probabilities for different instructions, and register live-in and live-out information.

3.3.3 Function-Level Reliability Models

Furthermore, in order to facilitate function/task-level reliability optimization, there is a need for reliability models at the corresponding granularity. Towards this end, two different reliability models are developed in this manuscript: (1) *Function*

Vulnerability Index (FVI), which is based on the IVI; and (2) *Function Resilience* model, which defines the resilience of a function as a probabilistic measure of the function's correctness (output quality) in the presence of hardware-level faults. The *function resilience* is a black-box model and relies on fault injection experiments to estimate the program reliability without exposing the software program details to the designers, and thus can also not be estimated statically. Besides this, to stay consistent with the IVI model, for the reliability optimizations and evaluations, the FVI metric will be used in the rest of manuscript. The details of the *function resilience* model and its applicability will be discussed in Appendix B. The function resilience model is beneficial in cases where designers do not want to dive into static analysis and in-depth program analysis.

3.3.4 Consideration for Timing Correctness

Besides functional correctness, this manuscript also accounts for timing reliability in timing-conscious systems. In order to account for the timing effects, this manuscript introduces a *Reliability-Timing Penalty* (RTP) model which is the linear combination of *functional reliability* (i.e., the reliability penalties in terms of the vulnerability indexes like FVI) and *timing reliability* (i.e., the timing penalty in terms of deadline misses). The RTP metric will be employed in run-time reliability optimization techniques in Chap. 6.

3.4 Consideration of Aging Faults

Besides soft errors, an on-chip system is also subjected to aging faults. Similar to the cross-layer reliability modeling flow for the soft error case presented in Fig. 3.8, a reliability modeling flow for the aging faults can also be obtained. There has recently been research going on in this area [116, 117]. Since this manuscript investigates soft error resilience in the presence of process variation and aging-induced effects, it requires input data on process variation maps and processor aging estimates for different workload scenarios. Towards this end, this manuscript developed a basic aging estimation technique (see details in Chap. 7) to validate the aging-aware soft error optimization concepts proposed in Chap. 6. The proposed aging estimation technique leverages accurate low-level aging estimates of different logic and memory elements obtained through detailed SPICE simulations (provided as an input by the VirTherm-3D group as a part of the collaborative research effort in the DFG SPP1500 program [118]). Besides this data, the technique requires the top $x\%$ critical paths from the circuit netlist and signal probabilities from the gate-level *ModelSim* simulations. The initial delay estimates are obtained from the *Standard Delay Files*, which are obtained as an output of the logical synthesis. The proposed technique generates aging estimates for $x\%$ critical paths by applying the delay degradation (obtained from low-level aging estimates) to individual logic elements in the *Standard Delay Files* and re-performs the critical path analysis. Finally, the aging of the processor's critical path is estimated as the accumulated delay of all logic elements considering their respective duty cycles over several years. Note that developing low-level aging models, consideration of temperature effects, thermal management, and thermal-aware processing are beyond the scope of this manuscript. Therefore, this manuscript considered aging estimates corresponding to the maximum operating temperature scenario. Details on the temperature-related optimizations can be found in works from other SPP1500 subprojects [149–151, 155].

The above-mentioned software program-level reliability models are leveraged to enable cross-layer reliability optimization across multiple system layers, i.e., at compiler and system software layers, that are briefly discussed in the following. Detailed discussions on the proposed techniques, experimental analysis, and results at different system layers are presented in Chaps. 5 and 6.

3.5   Reliability-Driven Compilation Flow

A reliability-driven compiler offers significant opportunities to generate reliable application code for unreliable or partially reliable hardware platforms. For effective reliability improvements, both *hardware knowledge* (e.g., number of available registers, area and fault probabilities of different processor components) and *software characteristics* (e.g., variable lifetime, instruction types and dependencies, basic block execution probabilities) need to be considered. The proposed reliability-driven compilation flow aims at improving the reliability of fault-susceptible software programs executing on unreliable hardware under user-provided tolerable performance overhead constraints. It improves the software program's reliability through two orthogonal and equally important ways: (1) Reducing the probabilities of Application Failure and Incorrect Output errors by reducing the spatial and temporal vulnerabilities and *critical instruction* executions. (2) Error detection and recovery through selective instruction protection.

Figure 3.10 presents the block diagram of the proposed reliability-driven compilation flow where the novel contributions of this manuscript are highlighted in the dark orange boxes, and the light orange represent the compile-time information that is exploited for static estimation of the reliability. The inputs to the compilation flow are: (1) application code; (2) application analysis for reliability and performance after profiling, for instance, the number of registers which are used and their vulnerable period in terms of lifetime; (3) the above-discussed software program-level reliability estimation models; and (4) the user-provided constraints for tolerable performance overhead. *Reliability-Driven Software Transformations, Instruction Scheduling,* and *Selective Instruction Protection* are proposed in the scope of this manuscript that can be applied in the front-/middle-/back-end of a reliability-driven compiler. The reliability-driven transformations and instruction scheduler lower the error probabilities by reducing the spatial/temporal vulnerabilities and the *critical*

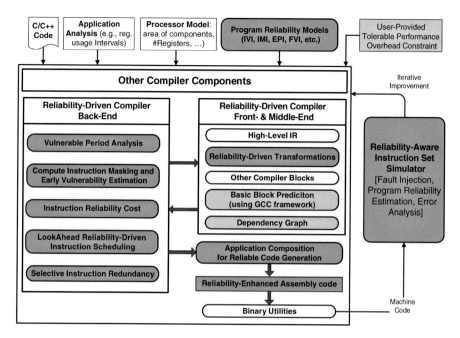

Fig. 3.10 Reliability-driven compilation framework

instruction executions. The selective instruction protection technique improves the software program's reliability through error detection and recovery. Using the above techniques, *multiple iso-functional code versions* are generated for different functions of an application software program, each exhibiting distinct performance and reliability properties. These multiple versions facilitate the run-time system to explore the reliability-performance optimization space at run-time considering varying fault rates and resilience/performance properties of the concurrently executing applications.

3.5.1 Reliability-Driven Software Transformations

The key goal of the reliability-driven transformations is to minimize the error probabilities towards Application Failures and Incorrect Outputs through the following two means under user-provided performance overhead constraints: (1) reducing the execution probability and vulnerabilities of *critical instructions* like branches, calls, load/stores, address generation and condition evaluation ALU instructions. (2) Reducing instructions' spatial and temporal vulnerabilities with respect to distinct usage patterns of various processor components (e.g., by reducing the register content and control flow vulnerabilities). In this manuscript, the following four reliability-driven transformations are proposed.

1. **Reliability-Driven Loop Unrolling** determines an "appropriate" unrolling factor that jointly minimizes the spatial and temporal vulnerabilities of different instructions under performance and code size constraints. It explores the tradeoff between reduced loop evaluation instructions and vulnerabilities of live register variables.
2. **Reliability-Driven Data-Type Optimization** targets reducing the number of *critical instruction* executions by transforming the smaller bit-width data types into larger bit-width data types for given data structures, while minimizing the function vulnerabilities.
3. **Reliability-Driven Common Expression Elimination and Operation Merging** reduces the vulnerabilities by removing identical expressions and/or merging partially common sub-expressions. It investigates the reliability effects (e.g., reduced Incorrect Outputs) of re-computation and register variables with increased lifetime while also accounting for the register spilling.
4. **Reliability-Driven Online Table Value Computation** evaluates whether pre-computed table values with increased memory vulnerability would be beneficial from the overall function reliability perspective or the online computation with increased instruction vulnerabilities in the pipeline.

When comparing to performance-optimized transformations, these reliability-driven transformations result in 60 % reduced Application Failures and on average 57 % reduced application vulnerabilities leading to reduced Incorrect Outputs; see detailed evaluation in Sect. 5.1.

3.5.2   Reliability-Driven Instruction Scheduling

A performance-driven scheduler may degrade the software reliability, for instance, by scheduling *critical instructions* after a pipeline stalling instruction or increased spatial vulnerability due to increased register usages. In this manuscript, a *reliability-driven instruction scheduling* technique is proposed that determines the instruction execution sequence under a user-provided tolerable performance overhead and directly influences the vulnerabilities of different instructions in different processor components, e.g., variable values in the register file or pipeline stage residency. The reliability-driven instruction scheduler improves the software reliability by prioritizing the instructions with the highest *reliability-weight*, which is a joint function of the statically estimated spatial and temporal vulnerabilities, instruction's criticality, probabilities of different types of software program errors, and number of dependent instructions. It employs a lookahead-based heuristic to evaluate the reliability weights of different scheduling candidate instructions in conjunction with their dependent instructions, thus minimizing the risk of scheduling a *critical instruction* after a multi-cycle or a pipeline stalling instruction (if possible). The proposed scheduler reduces Application Failures by 22 % on average compared to state-of-the-art instruction schedulers; see detailed evaluation in Sect. 5.2.

3.5.3   Reliability-Driven Selective Instruction Protection

Full-scale instruction redundancy techniques like [27] incur significant energy, code size/memory, and performance overhead, and thus cannot be applied in resource-constrained (embedded) systems. Moreover, different instructions in an application software program exhibit varying reliability importance due to their diverse error vulnerability and masking properties as a result of changing data and control flow behavior. For such scenarios, this manuscript introduces a *selective instruction protection technique* that applies redundancy to a subset of instructions with the highest reliability profit value under a given performance overhead constraint. The reliability profit is computed as a joint function of IVI, IMI, EPI, and protection overhead. In case of sequential dependencies, it may be beneficial to jointly protect a group of instructions as the voting and checking is only required at the end of the sequential group, thus incurring a reduced protection overhead. In constrained scenarios, the proposed selective instruction protection technique improves the software reliability by 30–60 % on average compared to different state-of-the-art techniques which do not jointly account for instruction-level vulnerability, error masking, error propagation, and hardware-level parameters; see detailed evaluation in Sect. 5.3.

3.5.4   Generating Multiple Function Versions Providing Tradeoff Between Performance and Reliability

The techniques explained in Sects. 3.5.1, 3.5.2, and 3.5.3 are leveraged to generate multiple compiled function versions that are identical in terms of their functionality and the output but differ w.r.t. their execution time and FVI properties. These versions can be generated in two ways: (1) after applying *only* reliability-driven transformations and instruction scheduling, such that these versions differ in their properties w.r.t. the reduced error probability; and (2) after *additionally* applying the selective instruction protection, so that these versions also exhibit some error detection and recovery. *These multiple function versions generated through reliability-driven compilation enable a reliability-performance optimization space for the system software layers.*

The number of different function versions would highly depend upon the range of tolerable performance overhead provided by the user. Note, the tolerable performance overhead per function can also be obtained from the system software layers after a slack analysis in a multi-pass cross-layer optimization flow. Furthermore, there may exist several non-optimal points in the reliability-performance optimization space of multiple versions. Therefore, in Sect. 5.4, function version generation and selection is discussed. The Pareto-optimal function versions in the reliability-performance optimization space are selected after applying different code-level reliability enhancing techniques. Furthermore, function versions with appropriate transformations will be selected in Sect. 5.1.6 to further curtail the design space and non-benefiting combinations of different transformations will be discarded. Curtailing the number of function versions is important to reduce the overhead of the contributions at the system software layers. *These multiple versions for different functions are then forwarded to the offline and run-time system software for further reliability optimization as discussed below.*

3.6   Reliability-Driven System Software

Figure 3.11 presents the overall flow for the reliability-driven system software which consists of the following two main steps: (1) reliability-driven offline system software; and (2) reliability-driven run-time system software. After multiple function versions are generated, the reliability-driven offline system software generates the offline function schedule for different versions, such that each schedule denotes a distinct combination of different function versions representing a distinct reliability and execution time profile for the complete function schedule for a given application. The offline prepared schedule aims at reducing the total system expected *Reliability-Timing Penalty* (RTP). These offline prepared tables of function schedules are given as an input to the reliability-driven run-time system software that manages dependable application execution on a single core or multi-/manycore processors.

The run-time system software for the single core processors chooses a certain function version from the appropriate schedule table depending upon the so far achieved system reliability and the remaining time to deadline considering the previously executed functions. Its goal is to minimize the total expected RTP for the complete function schedule. The run-time system software for the multi-/manycore processors performs three key operations: (1) it dynamically selects the redundant multithreading mode for different concurrently executing application tasks in order to improve their soft error resilience under resource-constrained scenarios. (2) Afterwards, it selects an appropriate set of cores for the tasks with or without redundant threads. (3) Then, it determines an appropriate compiled version for each of the tasks depending upon the cores' frequency, versions' execution time, and the deadline. This run-time system performs soft error resilience in the presence of design-time process variations and run-time aging-induced frequency degradations.

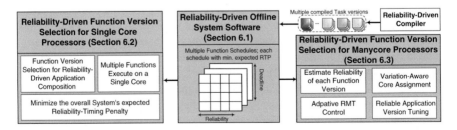

Fig. 3.11 Reliability-driven offline system software and run-time adaptive system software

In the following, both offline system software and run-time system software are briefly explained; see details on techniques and algorithms in Chap. 6.

3.6.1   Reliability-Driven Offline System Software

The proposed offline system software performs *reliability-driven application composition/function scheduling*. Given multiple reliable function versions generated from the reliability-driven compiler, each with a distinct reliability-performance tradeoff option, the reliability-driven offline system software constructs the scheduling table for all the functions using a dynamic programming approach. For different reliability and timing values, the RTP for all the function versions is estimated and the version with the minimum RTP is placed inside the table. The same process is done for all the functions and their corresponding versions depending upon the distributions of their execution time and vulnerabilities. A certain point across these tables presents a specific function schedule (as a solution for dependable application execution) that has a distinct combination of function versions improving the reliability in a certain execution context. The proposed technique constructs the schedule tables with the objective to minimize the overall RTP, therefore, the deadline misses versus the functional reliability tradeoff is exploited. Probability distribution of the task execution time is taken into account while constructing the schedule tables to exploit the dynamic execution behavior. The constructed tables of multiple function/task versions are given as an input to the run-time system software to dynamically select a certain schedule from the table that minimizes the overall system's RTP depending upon the current execution behavior (i.e., the reliability penalty and the remaining time to the deadline).

3.6.2   Reliability-Driven Adaptive Run-Time System Software

After generating the dependable code in form of multiple compiled versions and multiple function schedules, the proposed reliability-driven run-time system software manages the applications' executions considering the reliability/performance properties of different versions, the frequency variations in the underlying hardware, and the history of encountered errors. Towards this end, two different systems are proposed: (1) a reliability-driven run-time system software for the single core processors, and (2) an adaptive *Dependability Tuning* system for manycore processors subjected to core-to-core frequency variations as a result of process variations and aging-induced frequency degradation. The run-time system for the single core processors performs reliability-driven dynamic application composition by dynamically selecting an appropriate task version from the offline-constructed schedule tables depending upon the current execution behavior and reliability level while considering the remaining time to the final deadline. However, this run-time system software only targets soft error resilience in single core processors. When considering multi-/manycore processors, different cores in the on-chip system may exhibit different properties (e.g., operating frequency) due to design-time process variations and run-time aging. Therefore, the run-time system software for manycore processors needs to perform soft error resilience under process variations and aging-induced effects.

Towards this end, an adaptive *Dependability Tuning (dTune)* system for manycore processors is proposed in this manuscript. The *dTune* system performs the following key operations:

1. **Hybrid Redundant Multithreading Tuning**: Redundant threading is a well-adopted resiliency method in multi-/manycore processors, where redundant threads of different applications execute on different cores. In resource-constrained scenarios where multiple application tasks execute simultaneously, it may happen that there are not sufficient free cores for executing three redundant copies of each application task. In such cases, the resilience properties of different applications tasks and the history of encountered errors can be leveraged to selectively activate or deactivate redundant multithreading modes for these tasks.

2. **Variation-Aware Core Allocation**: Considering the existence of core-to-core frequency variations, a key challenge is to allocate cores to different redundant threads of different application tasks, such that the voting at the end of redundant multithreading can be performed with reduced synchronization overhead to ensure functional and timing reliability. For this, *dTune* employs a variation-aware core allocation strategy that finds cores with frequencies that match the execution properties of the redundant threads while considering the application tasks' deadlines.

3. **Reliable Code Version Selection**: Given multiple reliable code versions for different tasks, *dTune* selects an appropriate version considering the time to deadline and the frequency of the core.

Jointly considering multiple compiled versions and core-to-core frequency variations for soft error resilience enables our *dTune* system to achieve on average 44 % (maximum 63 %) improved system reliability compared to different state-of-the-art single-layer soft error optimization techniques, for different chip configurations with numerous variability maps at different aging years.

3.6.3 Comparing Cross-Layer vs. Single-Layer Reliability Optimizing Techniques

The proposed *dTune* system integrates the novel reliability optimizing techniques proposed in this manuscript to realize a cross-layer soft error optimization flow. Therefore, in order to illustrate the benefits of a cross-layer software reliability optimization over single-layer soft error optimization techniques, Chap. 7 will present a comprehensive comparative evaluation of the proposed *dTune* system for different chip configurations with numerous process variation maps and aging years. These experiments illustrate what the chip configurations and application scenarios are in which a cross-layer technique would be beneficial, and under which configurations and execution scenarios a single-layer reliability optimization technique is beneficial. In summary, the proposed cross-layer optimization flow is beneficial in most of the cases, but in particular, for the cases which are resource constrained.

3.7 Chapter Summary

This chapter presented the overview of the proposed cross-layer reliability analysis, modeling, and optimization flow and discussed the information exchange across multiple software layers. The necessary hardware knowledge is taken into consideration for software reliability modeling and optimization in order to bridge the gap between hardware and software to improve the effectiveness of the models and techniques. In order to devise efficient reliability models, a detailed software program-level reliability analysis is performed using fault injection experiments. The error distributions during the execution of different instructions are analyzed when faults are injected in different processor components. This study helps in understanding the impact of hardware-level faults at the software layer and their relationship to different instruction types and different processor components.

Also, different parameters for software-level reliability models are identified. Based on this analysis, instructions are classified into *critical instructions* (i.e., load, store, jump, call, and arithmetic/logical instructions used for address generation) and *noncritical instructions* (add, sub, and, or, etc.). Depending upon the usage of different processor components and time spent in these components by different instructions, the notion of spatial vulnerability and temporal vulnerability is introduced that define the space- and time-wise error probabilities, respectively. This analysis provides hints and motivation to consider different hardware and software parameters (i.e., area and fault probabilities for different processor components, spatial and temporal vulnerabilities, and the knowledge about the critical and non-critical instructions) for software-level reliability modeling and optimization. The message from this analysis is: the spatial and temporal vulnerabilities depend upon the instruction profile and processor architectural features. Therefore, both hardware and software knowledge need to be considered for reliability modeling at the software layers. Furthermore, different compiler and system software level techniques can affect the instruction profile and execution on a processor, respectively and thereby will affect the spatial/temporal vulnerabilities. Also, minimizing the number of critical instruction executions bears the potential of lowering the number of Application Failures that can be targeted through reliability-driven software transformations, as will be discussed in Chap. 5. Afterwards, an overview of the cross-layer reliability modeling flow is presented along with an introduction of the novel reliability metrics proposed in this manuscript; see details of these models (that are adapted to the granularity of appropriate system layers where these models are employed) and their evaluation in Chap. 4. Towards the end, novel software reliability optimization techniques at different system levels are proposed for dependable application code generation and execution. These techniques will be explained in detail in Chaps. 5 and 6.

Since this chapter only provided a high-level overview to connect all of the novel contributions of this manuscript in a cross-layer modeling and optimization flow, detailed discussion of results and achievements is presented in the Chaps. 4, 5, and 6, while the evaluation of benefits of cross-layer reliability optimization compared to the single-layer reliability optimization is discussed in Chap. 7.

Chapter 4
Software Program-Level Reliability Modeling and Estimation

In order to estimate the reliability at the software program- level while accounting for the knowledge from the underlying hardware layers, this chapter presents different reliability estimation models that are developed at different levels of granularity, i.e., instruction and function/task level. Since each system layer may employ distinct reliability optimization techniques that can operate at either the instruction or function/task level, it is important to devise reliability models for the appropriate granularity adapted to these optimization techniques. For example, a metric at an instruction granularity will be useful for enabling reliability optimization during compilation. However, at the system software layer the notion of function/task is more appropriate. A key challenge to develop efficient software program-level reliability models is to identify important hardware- and software-level parameters that affect the reliability of a software program executing on an unreliable hardware. For this, the analysis of Chap. 3 is important to be considered, i.e., the knowledge of critical and noncritical instructions, spatial and temporal vulnerability, and error masking can be leveraged to develop accurate software program-level reliability models. These models are then used to analyze the reliability properties of different applications at the instruction and function granularity.

In particular, this chapter first introduces three novel reliability models at the instruction granularity, namely Instruction Vulnerability Index, Instruction Error Masking Index, and Instruction Error Propagation Index that quantify three key reliability characteristics of the software program. The *Instruction Vulnerability Index* model (Sect. 4.1) estimates the probability of an instruction's output being erroneous. Quantifying instruction vulnerabilities poses several challenges to analyze and estimate (1) *spatial vulnerabilities* due to vulnerable bits of different operands of the instruction, area of different processor components, and their respective fault probabilities; and (2) *temporal vulnerabilities* due to the vulnerable residency of an instruction in different pipeline stages that also depends upon the previously executing instructions in the pipe-line. In case the output of an instruction is erroneous, the *Instruction Error Masking Index* model (Sect. 4.2) estimates the probability that this error will ultimately be masked until the final visible output of the application software program. This poses several challenges to characterize the masking potential of different types of instructions and analyzing the impact of other factors related to the data and control flow graph, for instance, control flow probabilities. In case the error is not masked, the *Instruction Error Propagation Index* model (Sect. 4.3) estimates that how many final outputs of the software program will be affected. This poses challenges related to control flow graph analysis for dif-ferent instruction paths until the program output. In order to employ these met-rics for devising reliability-driven techniques at the compiler and offline system software layers, it is also important to characterize these reliability properties (i.e., vulnerability, error masking, and propagation) *statically*. In summary, the key scientific challenges for all these models are fast estimation, parameter identification, impact, and estimation of hardware-/software-level parameters, etc. Furthermore, detailed analysis for different applications is performed and interesting observations are derived that are leveraged in the cross-layer reli-ability optimization flow. For instance, it is not necessary that an instruction with a high vulnerability also has a low masking probability. Also, modifying the vulnerabilities for some instructions during the optimization may have an impact on the vulnerabilities of the other instructions. Therefore, these program reliability models need to be carefully employed in the optimization flow.

Once these instruction-level models are obtained, at the function granular-ity, Function Vulnerability Index, Function Resilience, and Reliability-Timing Penalty models are proposed in Chap. 4 to quantify the error vulnerability, resilience to soft errors, and susceptibility to functional and timing errors, respectively. In particular, the *Function Vulnerability Index* model jointly accounts for the vulnerabilities of all instructions inside a given function, while the Function Resilience is more like a black-box model to provide a probabilis-tic measure of a function's correctness (in terms of amount of correct output) in the presence of hardware-level faults. Unlike the Function Vulnerability Index, the Function Resilience model does not expose the detailed low-level program analysis to the designers and is therefore beneficial in case a fast anal-ysis is required. Towards the end, the *Reliability-Timing Penalty* model is pre-sented that estimates the reliability of a software program as a joint function of its functional correctness and timing correctness. It is, in particular, beneficial for timing-conscious systems where both function and timing reliability are important.

The above-discussed models are discussed in the following sections in detail and will be leveraged in Chaps. 5 and 6 to design and evaluate different reliability optimization techniques for reliable code generation and execution.

4.1   Instruction Vulnerability Index

The *Instruction Vulnerability Index* (IVI) of an instruction i is defined as the accumulated vulnerability to soft errors during its execution in various pipeline stages using different processor components (Proc) while taking their respective area into account, for instance, in terms of vulnerable gates; see Eq. 4.1.

$$\text{IVI}_i = \frac{\displaystyle\sum_{c \in \text{Proc}} \text{IVI}_{ic} \times A_c \times P_{\text{fault}}(c)}{\displaystyle\sum_{c \in \text{Proc}} A_c} \qquad (4.1)$$

where c represents a certain processor component and A_c is its area (e.g., in terms of the number of gate equivalents). The parameter $P_{\text{fault}}(c)$ is the probability that a fault is observed at the output of component c. The parameter IVI_{ic} represents the individual vulnerability of the instruction i at a certain processor component c. It is defined as the normalized product of the *vulnerable period* (vulP_{ic}) and *vulnerable bits* (vulBits_{ic}) of a component c of an architecturally defined size TotalBits_c; see Eq. 4.2.

$$\text{IVI}_{ic} = \frac{\text{vulP}_{ic} \times \text{vulBits}_{ic}}{\displaystyle\sum_{c \in \text{Proc}} \left(\text{vulP}_{ic} \times \text{TotalBits}_c\right)} \qquad (4.2)$$

To bridge the gap between the hardware and software, the IVI model considers the knowledge from both the software and the hardware layers. **The software-level knowledge** is incorporated through *vulnerable period* and *vulnerable bits* in the IVI_{ic}, which are the ISA-visible state to the application software program. The parameter vulP_{ic} denotes the *temporal vulnerability*, whereas the parameters vulBits_{ic} and A_c denote the *spatial vulnerability*. The vulBits_{ic} is obtained by performing comprehensive software-level vulnerable bit analysis that captures the vulnerable portions of the architectural components by exploiting the read/write dependencies w.r.t. the register file and instruction dependencies, i.e., without considering fault injection and the underlying microarchitecture details. For instance, in case of a register, the bits written in a certain cycle but not read are denoted as *non-vulnerable bits*. **For hardware-level knowledge**, the parameter A_c and $P_{\text{fault}}(c)$ are considered in the total vulnerability estimation in Eq. 4.1. The number of particle strikes depends upon the area of a component, for instance, a 32-bit register and a multiplier will experience different number of particle strikes over the period of time. To incorporate this factors, the parameter A_c is employed as it captures the actual physical area-wise *spatial vulnerability* and it is obtained through the hardware synthesis results using *Synopsys Design Compiler*. The parameter $P_{\text{fault}}(c)$ is employed to incorporate the *logical masking* effects. The value of $P_{\text{fault}}(c)$ depends upon the microarchitecture of the component c, i.e., how different types of gates are connected in its circuit. It is obtained after performing a detailed gate-level analysis. Note, in case of the register file, the $P_{fault}(register\ file)$ is considered 100 % because a fault in the register will be latched until rewritten. Furthermore, in case a processor component is fully protected using some hardware-level protection mechanism like ECC or TMR, then its $P_{fault}(c)$ is considered to be 0. For design-time/compile-time optimizations, the IVI and the corresponding parameters, i.e., *vulnerable period* and *vulnerable bits* are statically estimated. For evaluation, these parameters and IVI are obtained using the execution trace of an application software program. In the following sections, the estimation of the above-mentioned parameters is explained.

4.1.1   Estimation of Vulnerable Periods

The vulnerable period can be estimated both *statically* and *dynamically* depending upon how the IVI is estimated. For static IVI estimation, the vulnerable period is determined using the compile-time knowledge like control and data flow graph (CDFG) with basic block execution probabilities, and instructions' latencies in different processor components considering 100 % cache hit. However, for dynamic estimation of IVI, the run-time knowledge like the actual vulnerable periods of an instruction for a given software program's execution trace are considered. In the following, both methods are discussed for the register file and pipeline components.

Vulnerable Period Estimation for the Pipeline Components: The CDFG presents the instruction dependency graph which presents the execution sequence of instructions inside the pipeline. This gives an idea of the vulnerable period of an instruction inside the pipeline since the vulnerable period of an instruction in different pipeline stages depends upon the residency time of instructions in the succeeding pipeline stages (i.e., the previous instruction in the execution flow). Therefore, the vulnerable period highly depends upon (1) execution latency of the instructions, e.g., multi-cycle instructions such as *multiply* and *divide* instructions that may stall the pipeline for several cycle in the execute stage[1]; and (2) the instruction schedule. In case a multi-cycle instruction stalls the pipeline during the execution (e.g., a *multiply* instruction), the temporal vulnerability of the instruction in the preceding pipeline stage (i.e., the decode stage) is increased. Statically, the vulnerable period of an instruction is the residency time (or in other words the occupancy cycle) of an instruction inside a certain pipeline stage. A 100 % cache hit rate is considered such that the vulnerable periods of instructions can be estimated in a predictable way, i.e., the time that each instruction takes inside a certain pipeline stage, for instance, a *multiply* instruction is considered take three cycles in the execute stage and a *load/store* instruction is con-

---

[1] Furthermore, multiplier and divider exhibit higher spatial vulnerability due to their increased area. Temporal and spatial vulnerabilities of these functional units depend upon the microarchitecture.

sidered to take two cycles. The instruction types and their execution sequence are known during compile time from the CDFG. Therefore, the vulnerable period for all instructions in different processor components can be statically estimated using the CDFG and the architectural model of the pipeline. For run-time estimation of IVI, actual execution trace needs to be taken into account because the vulnerable periods will vary depending upon the branches and the cache misses. For example, latencies due to the cache miss in the memory stage will result in an increase in the vulnerable periods of all the instructions in the preceding pipeline stages.

Vulnerable Period Estimation for the Register File: The vulnerable period of an instruction w.r.t. the register file depends upon the number of operand variables (stored in registers) which are used by an instruction and the time when these registers are *set* and are *used*. Figure 4.1 demonstrates an example scenario illustrating the procedure of computing the vulnerable periods for operand variables stored in registers. The two operand variables of an instruction have different vulnerable periods (i.e., lifetime of the operand variables in terms of cycles). For an instruction, the vulnerable periods of its operand variables depend upon the latency of previously executed instructions and instruction dependencies because a pipeline stalling instruction increases the vulnerable period of the operand variables. For example, for the fifth instruction at cycle #9, the vulnerable periods for *R0* and *R2* are four and six cycles, respectively. In case the operand variables are alive from the previous basic block, besides the previously scheduled instructions (i.e., predecessor instruction in CDFG), static estimation of the vulnerable periods w.r.t. the register file also depends upon the execution probability of the previous basic blocks.

Figure 4.2 illustrates the computation of vulnerable periods (VP_{R1} and VP_{R2}) of two operand variables (R_1 and R_2) of an *add* instruction in the basic block B_4. This instruction uses the variable values which are alive from the previous basic blocks B_1, B_2, and B_3. When following the control flow through basic block B_2, the vulnerable period of R_1 (i.e., VP_{12}) is less than that of R_2 (i.e., VP_{22}), as the value of R_1 is overwritten in the basic block B_2. In contrast, the vulnerable period of R_2 (i.e., VP_{21}) is less than that of R_1 (i.e., VP_{11}) in case the control flow passes through B_3 as the value of R_2 is overwritten in the basic block B_3. This demonstrates that, in case an

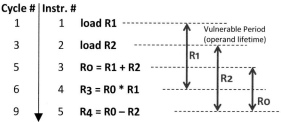

VulnerablePeriod = (Cycle of Current Usage − Last Write Cycle)
for i= 5; VulnerablePeriod$_{R0}$ = 4 Cycles; VulnerablePeriod$_{R2}$ = 6 Cycles

Fig. 4.1 An abstract example illustrating the vulnerable periods of operands

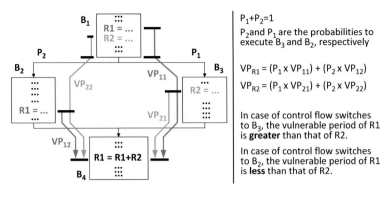

Fig. 4.2 Computing the vulnerable periods for operand variables of the ADD instruction in the B_4 basic block under changing control flow

instruction uses the operand variables alive from previous basic blocks, the effective vulnerable period depends upon the execution probability of the previous basic blocks. Therefore, for static IVI estimation, the vulnerable periods of the two operand variables R_1 and R_2 are scaled according to the basic block execution probabilities relative to other previous basic blocks (see computation of VP_{R1} and VP_{R2}). In this manuscript, the branch prediction feature of the GCC framework is employed to obtain the execution probability of each basic block relative to the previous basic blocks. This feature is activated using option *-fguess-branch-prob* and it uses different heuristics [121]. For every branch, a probability is estimated that it will be taken.

The dynamic estimation of IVI w.r.t. register file requires the execution trace analysis for *set* and *use* of different registers used for storing operand variables, i.e., for how much time each operand variable is alive inside a register and when is its value (re-)written. This will vary depending upon the input data for the application software program.

4.1.2 Estimation of Vulnerable Bits

It is important to consider that not all bits of operand variables are vulnerable for the correct software execution due to inter-layer masking from microarchitecture-state to the ISA-visible state [31, 74] (i.e., a fault in the hardware does not necessarily lead to an erroneous output after the instruction execution). If a bit is necessary to stay correct for an error-free instruction execution (i.e., a fault in this bit will corrupt the output of the instruction), it is classified as a *vulnerable bit*. All other bits are *non-vulnerable bits*. To demonstrate this effect, Fig. 4.3 presents an example code of a *bitwise and* instruction with operand variables R_1 and R_2. This instruction extracts the lower 16 bits of R_1. Let us assume that a fault may occur either in R_1 or R_2, but it is less probable that a fault can occur in both. If a fault happens in the upper 16 bits of R_1 (Fault Scenario-1), this will not affect the value of R_3. However, if a

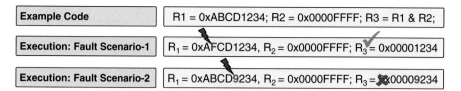

Fig. 4.3 Example for vulnerable bits computation of the for operand variables of the AND instruction

fault occurs in any bit of R_2 or in the lower 16 bits of R_1 (Fault Scenario-2), this will affect the value of R_3. Therefore, bits 0–15 of R_1 are *vulnerable bits*, while bits 16–31 of R_1 are *non-vulnerable bits*. In contrast to this, all 32 bits of R_2 are vulnerable. So in total out of 64 bits (i.e., R_1 and R_2), 48 bits are vulnerable whereas 16 bits are non-vulnerable.

For an application software program, a potential method for estimating the vulnerable bits is fault injection experiments. However, as the number of instructions grows, it may not be possible to inject the faults in every bit of the operand variables for every instruction and analyze its effects at the corresponding instruction's output in a realistic time. Moreover, it is extremely time consuming to do this analysis when considering different input stimuli for the software program because as soon as the input values change, the number of vulnerable bits will also vary. To alleviate this analysis overhead, another method is to statically estimate vulnerable bits, as adopted in this work. Given one or both of the input operands for an instruction are constant value(s), the vulnerable bits are estimated w.r.t. the constant value and the instruction type, which can be obtained from the assembly code or CDFG, as explained above with the help of an example in Fig. 4.3. In case input operands are variables, all 32 bits are considered vulnerable for the ease of evaluation and static estimation. It is important to note that for static estimation, considering all 32 bits of operand variables is anyway the solution because the exact values of these variables are unknown at the compile time. Further, the IVI model is independent of this simplification and any complex/comprehensive vulnerable bit analysis technique can be employed to achieve even more accuracy.

4.1.3   Estimation of Component-Level Fault Probabilities

In this section, the procedure of estimating the $P_{fault}(c)$ for different processor components is shown. Figure 4.4 illustrates an example of an instruction i executing through different pipeline stages/components (PC) and corresponding masking probabilities $P_{EM}(i, PC)$, s.t., $PC = \{F, D, E, M, W\}$. The microarchitecture-level logical masking for an adder circuit (which is a part of the ALU) is shown as an example. An error at the "AND" gate is blocked/masked by the subsequent "OR" gate (see Case-1). However, the error at the "XOR" gate is propagated to the output (see Case-2). It is noteworthy that only "OR" and "AND" gates can mask the error

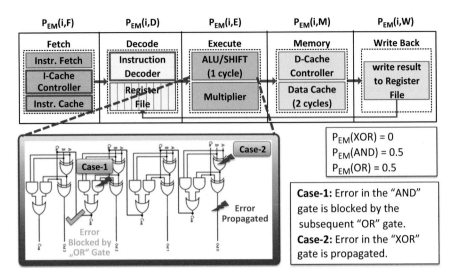

Fig. 4.4 Different pipeline stages exhibit distinct masking during the instruction execution due to combinatorial logic

with a theoretical average-case probability of 0.5 (the masking probabilities change depending upon the inputs of the logic gates). However, "XOR" and "NOT" gates do not mask an error. Once the masking probability is obtained, the $P_{\text{fault}}(c)$ can be computed using Eq. 4.3.

$$P_{\text{fault}}(c) = 1 - P_{\text{EM}}(c) \tag{4.3}$$

The parameter $P_{\text{EM}}(c)$ represents the error masking probability in pipeline component c due to microarchitecture-level logical masking effects, i.e., the error within a pipeline component (combinatorial logic) is not visible at the output latch as the error propagation through different logic elements/gates is blocked due to subsequent logic element(s). Consequently, the output of the pipeline component remains correct. The error masking probability in pipeline components can be obtained in two ways: (1) gate-level error propagation analysis of the netlist of a given processor component, (2) through extensive fault injection experiments. The logical masking properties of a pipeline component depends upon the structure of the microarchitecture; hence, the logical masking properties of, for instance, a carry-lookahead adder are different from a ripple-carry adder.

To estimate the fault probability, first, an error masking analysis is performed for each *path* in the component. A *path* is defined as a series of connected gates in a component which starts with an input gate and ends with an output gate such that, at each step there is only one gate. This facilitates independent analysis of all paths while still considering the error propagation across different paths. The inputs to the error masking analysis are: circuit netlist, particle strike-to-error probabilities for logic and memory elements in the netlist, and signal probabilities.

For a logic gate or memory element, particle strike-to-error probability is typically denoted as logic element error probability or bit error probability, respectively. These error probabilities are typically vendor specific and depend upon various technology parameters [113] and are typically unavailable under an open-access program. Finding these probabilities typically require detailed SPICE-level simulations and can be done using the Predictive Technology Model [114] data. Since this is beyond the scope of this manuscript, without the loss of generality, in this manuscript these probabilities are considered as 1 due to the unavailability of this technology data. This somehow leads to a more pessimistic estimate of component-level fault probabilities. The signal probabilities are obtained through gate-level simulations using *ModelSim* [124] software. It is important to note that not all faults in a logic element (i.e., gates) of a combinatorial circuit affect its output because a fault may be masked due to the logical masking effects of the subsequent gates. The fault probability for a processor component can be estimated using the average-case fault probabilities of different gates in its circuit netlist and performing a conditional probability analysis of error propagation through multiple gates until the circuit output using the following procedure.

At first, the masking probabilities of individual gates in the netlist are estimated. Afterwards, the error masking probability at a gate along a path (i.e., a set of connected gates leading to an output) is estimated. It is given as the probability that an error at this gate will not affect the correct output of this path, and it is calculated based upon the gate input signal probabilities and error probabilities. Following are the number of steps for calculating the error masking of a component:

1. **Finding All Paths**: The first step is to extract all the paths in the netlist of a given processor component, which is obtained after the processor is synthesized. A depth first search algorithm is implemented which takes a graph of nodes and edges as an input, where nodes and edges represent the gates and their connections. The algorithm traverses all the paths starting from the first node (i.e., which has no predecessor node that passes an input) and ending with the node which has no successor to which it outputs. In case of loops, the algorithm only takes the path for one iteration. Once all the paths branching from the same start gate are found, the algorithm will move to the next starting gate and finds all the paths which emerge from that starting gate.

2. **Gate-Level Error Masking Calculation**: After the paths are extracted, the masking probabilities are computed at the gate level. The masking calculations for different types of gates are different and are computed using the input signal probabilities and input error probabilities. For example, for a two-input AND gate, the signal probabilities for both input signals are assumed to be 0.5, denoting that the two input state signal probability (i.e., "1" and "0") is 50 %. However, in case of the output of a two-input AND gate, the state signal will be "1" only when both the input signals are "1" else it will be "0," which means that the total output probability for signal "1" is 0.25, whereas for signal "0" its 0.75. See the output signal probability computation below in Eq. 4.4.

$$P_{\text{sig}(0)_{A=0,B=0}} = 0.5 \times 0.5 = 0.25; \quad P_{\text{sig}(0)_{A=0,B=1}} = 0.5 \times 0.5 = 0.25$$

$$P_{\text{sig}(0)_{A=1,B=0}} = 0.5 \times 0.5 = 0.25$$

$$P_{\text{sig}(0)} = P_{\text{sig}(0)_{A=0,B=0}} + P_{\text{sig}(0)_{A=0,B=1}} + P_{\text{sig}(0)_{A=1,B=0}} = 0.75$$

$$P_{\text{sig}(1)_{A=1,B=1}} = 0.5 \times 0.5 = 0.25$$

(4.4)

If an error occurs at any of the input signals, it is important to analyze how it will affect the final output. Now, let us assume that two inputs are 01 and a bit flip occurs at the input "0," this will result in a corrupted output value "1" instead of "0" which is the correct value. There are four erroneous states out of the eight total numbers of states; therefore, the basic error masking probability of a two-input AND gate is as $P_{\text{EM gate}} = \dfrac{4}{8} = 0.5$. Finally, after computing the input signal probabilities and multiplying with the error masking probability (0.5), the gate-level error masking probability is computed using Eq. 4.5.

$$P_{\text{EM}} = 1 - \left(\begin{array}{c} P_{\text{EM Gate}} \times P_{\text{sig}(0)_{A=0,B=1}} + P_{\text{EM Gate}} \times P_{\text{sig}(0)_{A=1,B=0}} \\ + P_{\text{EM Gate}} \times P_{\text{sig}(1)_{A=1,B=1}} \end{array} \right) = 0.625$$

(4.5)

3. **Path-Level Error Masking Calculation**: Once the error masking probability for each gate is obtained, the masking probability for the complete logical path in the circuit is computed staring from the last gate of the path, i.e., in the reverse direction. The masking probability of a path is: if a soft error occurs at the first gate of the path, the probability that the error is masked until the final circuit output. After finding the masking probabilities for all the paths, the final masking probability at the circuit output $P_{\text{EM}}(c)$ for a component c is obtain as the product of the masking probabilities of all the paths. The probability values tell how many errors are masked and how many are propagated at the final circuit output. For the complete path, the calculation of the masking probability is very time intensive, since the paths in the graph of a complex circuit could be long enough. To minimize the computation time, the idea of reverse order masking computation is employed. An approach to speed up this analysis is proposed in [115]. Therefore, the masking probability is computed starting from the last gate.
4. **Fault Probability**: The component level fault probability, i.e., $P_{\text{fault}}(c)$ for a component c, is obtained by subtracting $P_{\text{EM}}(c)$ from 1.

$$P_{\text{fault}}(c) = 1 - P_{\text{EM}}(c)$$

(4.6)

4.1.4   IVI Results for Different Applications

Figure 4.5 shows varying distribution for the instruction vulnerabilities for different instructions; horizontal axis shows the instruction ID (from the assembly code) and the vertical axis shows the IVI value. The variations in the instruction vulnerabilities are due

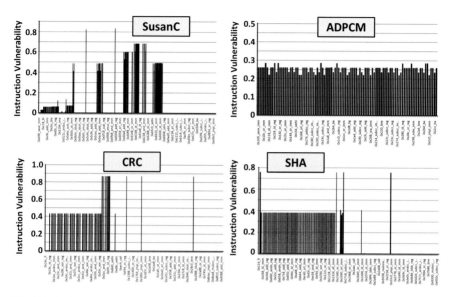

Fig. 4.5 Distribution of instruction vulnerability index for different instructions inside each application

to the unique instruction profiles of different applications and the corresponding variations in instructions' vulnerable periods in different pipeline components. For example, when comparing SusanC with ADPCM, it is observed that the instruction vulnerabilities in SusanC are relatively higher compared to that of ADPCM. Though SusanC has a relatively sparse distribution, the reason for high vulnerability of its instructions (even close to 1) is: several instructions are *either* dependent upon the global register variables that may be alive throughout the lifetime of a software program *or* their predecessor instructions were executed much earlier that led to an increased lifetime and consequently higher temporal vulnerability of dependent operand variables. Furthermore, even considering 100 % cache hits, there are more pipeline stalling multi-cycle instructions in SusanC (e.g., *multiply* and *load/store* instructions) which tend to increase the vulnerable periods for the instructions in the preceding pipeline stage. This would even worsen in the presence of pipeline stalls due to cache misses. Similar reasons also hold for other applications and result in varying IVI distributions.

In case of SHA, the reason for the instruction with similar vulnerability values is: these instructions experience similar execution flow determining similar temporal vulnerabilities, e.g., operating on a new register value with typically a vulnerable period of one cycle, instructions with a single cycle execution latency without any stalls, etc. However, the instructions with high vulnerabilities are the ones which are operating on the register values that are alive for a longer time and they are scheduled after the pipeline stalling instructions like a multi-cycle instruction. Similar behavior can also be noticed for the CRC and ADPCM applications.

Summary and Observations from the Above Results: The variation in the vulnerability distribution of different applications and instructions hints towards their reliability-wise higher or lower impacts. For example, SusanC has some instructions which have higher vulnerability, and therefore these instructions are more important for correct execution and need to be protected first, whereas in ADPCM, the distribution is smooth; therefore, all the instructions have equal reliability-wise importance and are required to be protected. For a given tolerable performance overhead, an instruction with a higher vulnerability would be a good candidate for protection first.

However, *Instruction Vulnerability Index* only covers one reliability aspect for software programs. It may happen that an instruction with a high vulnerability (i.e., with a high error probability) produces an erroneous output which is later masked due to subsequent instructions and control flow. Similarly, an instruction with a relatively low vulnerability may have a high error propagation probability to the final output of the software program. Therefore, it is important to quantify the error masking and propagation effects. In the following sections, two novel models *Instruction Error Masking Index* and *Instruction Error Propagation Index* are presented that quantify the error masking probability and error propagation effects at the instruction granularity.

4.2 Instruction Error Masking Index

The program-level error masking is quantified as the *Instruction Error Masking Index* IMI(I) of an instruction I, which is defined as the total probability of an error at instruction I being masked until the last instruction of all of its instruction paths, p (i.e., leaf nodes), such that the output of p is correct.

4.2.1 Parameter Identification

The masking of an error at an instruction I can happen because of the following factors that constitute the parameters of the IMI(I) model.

1. **Error Masking Due to Data Flow Properties—$P_{DP}(I, p)$**: The error at an instruction I can be masked because of the successor instruction in path p depending upon two factors: (a) the type of an instruction, and (b) the value of the operand variables.
2. **Error Masking Due to the Changing Control Flow Properties**: On the highly probable execution path, there may be masking instructions that can block the propagation of the error to the relevant output of the software program. It can also happen that a highly probable execution path does not even use the erroneous output value from the preceding basic block(s).

Figure 4.6 provides an overview and flow of steps to estimate the error masking probabilities at instruction level (for the ease of reader, Appendix D provides the complete list of notations/symbols).

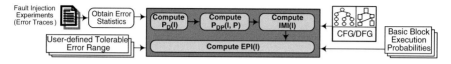

Fig. 4.6 Flow of steps to compute the instruction-level masking probabilities and error propagation index

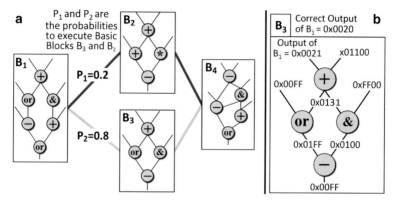

Fig. 4.7 An example control and data flow graph showing the error masking due to successor instructions and changing control flow

4.2.2   An Example

An example of error masking due to the control flow properties is presented in Fig. 4.7, showing different basic blocks and the control flow probabilities. The chances for the errors in basic block B_1 to be blocked in B_3 are higher due to the "&" and/or "or" instructions. However, if the control flow follows B_2, the chances for the error to propagate to B_4 are higher, which will ultimately get visible at the output of the software program. Note that the execution probability of basic block B_3 is higher compared to the basic block B_2. Only in case of the "&" and "or" instructions, the error masking occurs because of the value of the other operand, while "+" does not mask the error.

4.2.3   Parameter Estimation

From the software program's perspective the most important parameters that need to be estimated are the masking probabilities $P_{DP}(I,p)$ and IMI(I), as discussed in the following.

Estimation of $P_{DP}(I, p)$: The masking probability for each instruction I, depending upon the data flow can be modeled using Eq. 4.7.

$$P_{\mathrm{D}}(I) = \sum_{x \in I.O.Bits} P_{\mathrm{D}}(x,I) \times P_{\mathrm{e}}(x) \qquad (4.7)$$

where O is the set of operands of instruction I with a set of Bits. The masking probability of each bit is represented by the parameter $P_{\mathrm{D}}(x, I)$ which depends upon the instruction type. The parameter $P_{\mathrm{e}}(x)$ is the error probability of each bit, which can be simplified to $1/N_{\mathrm{Bits}}$ by assuming the same error probability for all bits, where N_{Bits} is the bit-width of the operand registers. In case a tolerable range th is specified by the user for the error in the output value ($\log_2(th)$ provides the number of bits), Eq. 4.7 can be modified as Eq. 4.8.

$$P_{\mathrm{D}}(I) = \sum_{x \in I.O.Bits \backslash \log_2(th)} P_{\mathrm{D}}(x,I) \times P_{\mathrm{e}}(x) + \log_2(th)/N_{\mathrm{Bits}} \qquad (4.8)$$

In a given instruction path p, the error masking depends upon both the individual instructions and the combination of consecutive instructions. Depending upon the masking behavior, instructions can be classified into two categories (explained below with the help of examples in Fig. 4.8).

Type A: Instructions such as "&" and "or" with a variable value, assuming a random bit masking, have a theoretical average-case masking probability of 0.5 (i.e., in 50 % of the cases the error is masked). In case of two consecutive "&" instructions in an instruction path, the total masking probability along the path P_{DP} is 0.75 (i.e., in 75 % of the cases the errors are masked), as the first "&" instruction will mask 50 % of the errors, and the successive "&" instruction will mask the remaining errors by 50 %.

Type B: For the instructions like "shift left" or "shift right" by any constant value (such that the masking bits can be inferred from the CDFG or assembly code), the computation of masking probabilities for the predecessor instruction is affected in a different way. If there are two consecutive shift instructions with a shift amount of 16 bits in an instruction path (e.g., "sll"), the total masking probability along the path P_{DP} is 1 (i.e., 100 % errors are masked), as after two consecutive 16 bit shift instructions all the 32 bits are shifted out. This observation hints that different combination

Fig. 4.8 Impact of different instruction types on the total masking probability along the instruction path

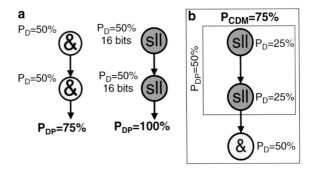

of the instruction types can have a different influence on the total masking probability along the instruction path $P_{DP}(I, p)$. If the joint masking effects of the consecutive instructions are ignored, the resultant masking probabilities are underestimated.

In the following, the step-by-step process for computing the masking probabilities is explained with the help of an example.

An Example Illustrating the Process of Computing Masking Probabilities for Consecutive Instructions: An example scenario is presented in Fig. 4.9, that illustrates the computation of error masking probabilities for different instructions in a given instruction graph, while showing how consecutive instructions of type B affect the total masking probability. It is shown that, when considering the masking effects of consecutive instructions of type B (i.e., instruction-1 and instruction-2), the total masking probability for instruction-1 becomes equal to "0.803." However, if the masking effects of consecutive instructions are ignored, the total masking probability of instruction-1 would be equal to "0.770," i.e., a difference of "0.033" compared to the earlier case. It is evident from this example that ignoring the effects of consecutive instructions may lead to an underestimation of masking probabilities. Hence, for an accurate estimation of instruction-level error masking, it is imperative to account for the joint error masking effects due to the combination of consecutive successor instructions of type B in the path p.

Computing the $P_{DP}(I, p)$: Figure 4.10 presents the flow of computing the error masking probability $P_{DP}(I, p)$ for instruction I of path p in an iterative manner starting from the leaf nodes (see detailed algorithm in Appendix C). The inputs are the function graph $G(V, E)$, set of leaf nodes L_G, set of predecessors and successors for each instruction (P, S). First, for all instructions in G, the error masking probability $P_D(I)$ is computed using Eqs. 4.7 and 4.8. Afterwards, $P_{DP}(I, p)$ is initialized for all the leaf nodes with their corresponding $P_D(I)$, as they represent the last nodes of the graph. The list L of ready nodes (i.e., instructions for which $P_{DP}(I, p)$ can be computed) is initialized with the predecessors of leaf nodes. The process to compute the $P_{DP}(I, p)$ is iterated until the list is empty. In each iteration, first, the set of all possible instruction paths is generated from every instruction

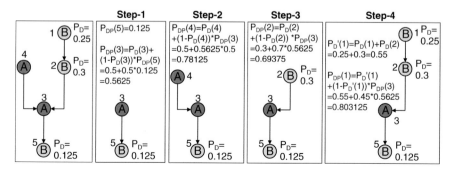

Fig. 4.9 An example showing the computation of error masking probabilities illustrating the effect of consecutive instructions of type B in the path on the total masking probability

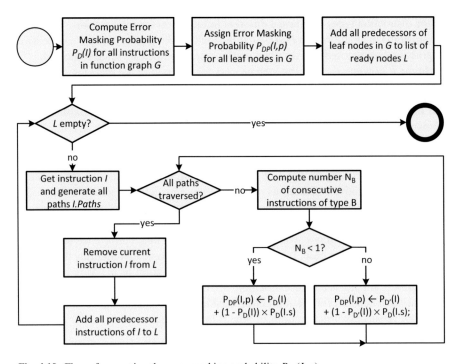

Fig. 4.10 Flow of computing the error masking probability $P_{DP}(I, p)$

until the leaf nodes. Afterwards, for all possible paths of every instruction, the number of consecutive instructions of *type B* (N_B) is computed and the masking probabilities $P_{DP}(I, p)$ are computed. If there are two or more consecutive instructions of *type B*, their cumulative masking probabilities are considered to compute $P_{DP}(I, p)$; otherwise, the independent masking probabilities are considered. Afterwards, the instruction under evaluation is removed from the list and its predecessors are added to the list.

Estimation of IMI(*I*): The *Instruction Error Masking Index* IMI(*I*) for a given instruction *I* is modeled using Eq. 4.9 as the error masking probability due to both data and control flow. IMI(*I*) of an instruction *I* can be modeled as the weighted masking probability due to data flow $P_{DP}(s, p)$ for all the *successor* instructions *s* of instruction *I* and in all the corresponding instruction paths *s.Paths* while accounting for the execution path probabilities $P_{CF}(ep|I)$; see Eq. 4.9.

$$\text{IMI}(I) = \sum_{\forall ep|I \in ep} \left(P_{CF}(ep|I) \times \prod_{\forall s \in I.S; \forall p \in s.Paths} P_{DP}(s,p) \right) \qquad (4.9)$$

The execution path probabilities $P_{CF}(ep|I)$ are estimated using the branch prediction functions of the GCC framework through the option "*-fguess-branch-prob.*"

The IMI(*I*) is computed using the *breadth-first search starting from the leaf nodes* and explores *G* in a bottom-up fashion.

4.2.4 Instruction Error Masking Index Results for Different Applications

Figure 4.11 shows the distribution of masking probabilities for SusanC, ADPCM, DCT, and SAD applications. The horizontal axis shows the instruction address in the execution sequence and the vertical axis shows the IMI value. It is important to note that for many instructions the IMI values are close to 0, 0.5, and ≈1. This primarily reflects the following three important cases.

1. **Case-1 with IMI Close to 1**: An instruction is reliability-wise not important because almost all errors in its output will be masked before the final output of this software program, thus an error in the output of this instruction will not matter. This happens mostly for instructions that have several successor instructions with very high masking probabilities.
2. **Case-2 with IMI Close to 0.5**: This indicates the cases where there are a few successor instructions like comparison or logical instructions that have a masking probability of 0.5.
3. **Case-3 with IMI=0**: This indicates the cases where either there is no control flow or an instruction does not have any error masking successor instruction in all the possible paths from this instruction until the final output. An example

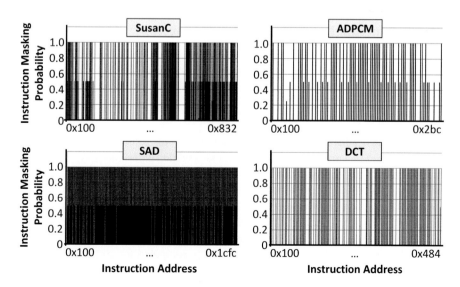

Fig. 4.11 Distribution for the instruction error masking indexes

could be a sequence of arithmetic instructions leading to an output value as in the case of DCT application for many instructions.

In Fig. 4.11, in case of SAD application, the case-2 is dominant due to comparison instructions of the *absolute* operation. However, case-3 is dominant in DCT due to dependent arithmetic instructions. The SAD application has a homogeneous distribution showing that many instructions can mask the errors due to the existence of more logic operations. The SusanC application exhibits the instructions which have a higher error masking potential and the reason is that SusanC has concatenated branch instructions and if an error occurs in the first branch, the next branch instruction masks the error. The ADPCM application has a relatively less number of logical instructions (i.e., "&" and "or") that can mask the errors. There are more non masking instructions (e.g., arithmetic, call, jump, and branch) which once affected by the errors, propagate this error to the output.

Summary and Observations from the Above Results: Besides consideration of *Instruction Vulnerability Index*, *Instruction Error Masking Index* is also important to be taken into consideration because several erroneous outputs can be masked in the CDFG. In case the errors are not masked, these *unmasked* errors may propagate to their successor instructions in different execution paths and can potentially corrupt several output values. In several application cases, error propagation to multiple execution paths and multiple outputs will be considered as a more reliability-worse situation. Furthermore, for fault containment, reducing the error propagation is also important. In the following, a novel *Instruction Error Propagation Index* model is proposed that quantifies the error propagation effects for different instructions in the CDFG, which provides a metric for the severity of the unmasked errors.

4.3 Instruction Error Propagation Index

Definition: The *Instruction Error Propagation Index* EPI(I) is defined as the product of the non-masking probability (i.e., the probability that an error is *visible* at the program output) of all the *successor* instructions of a given instruction I for all possible instruction paths.

Estimation Procedure: Figure 4.12 shows the flow for estimating the EPI(I) for each instruction in a given function graph G (see detailed algorithm in Appendix C). First, the EPI for all the leaf nodes (outputs) is initialized with 1, as the errors in the leaf nodes are considered as propagated to the next stages of the program execution. A list C is used in order to track the traversed instructions for which EPI is computed. Moreover, a FIFO-based queue Q is used to store the instructions for which EPI can be computed considering all successors are completed. Initially, the predecessors of the leaf nodes are inserted in Q and the EPI(I) is computed for all the instructions whose successors are in the C list. Otherwise, the instruction I is inserted back into the queue Q. For EPI(I) computation for an instruction I, all of its successor's instructions $I.S$ and their corresponding non-masking probabilities are considered. Note that

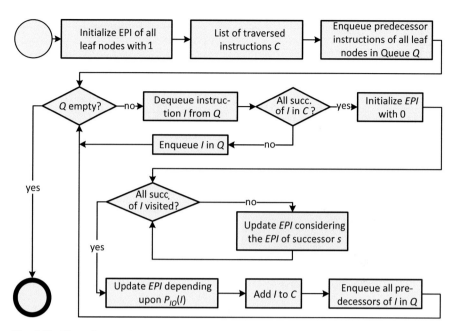

Fig. 4.12 Flow of computing the instruction error propagation index

the EPI(I) computation accounts for the probability of Application Failures $P_{\mathrm{AF}}(I)$ and probability of Incorrect Output $P_{\mathrm{IO}}(I)$; see Eqs. 4.10 and 4.11. COB(I) is the number of critical opcode bits that lead to a "non-decodable instruction" error. $P_{\mathrm{eOP}}(b, I)$ and $P_{\mathrm{eAd}}(b, I)$ are the error probabilities in the opcode and address bits. CAB(I) is the number of critical address bits that lead to a "memory segmentation" error due to an access to the invalid or restricted memory region. After computing the EPI(I) for instruction I, all of its predecessors $I.P$ are added into the queue Q.

$$P_{\mathrm{AF}}\left(I\right) = \sum_{b \in \mathrm{COB}(I)} P_{\mathrm{eOP}}\left(b,I\right) + \left(1 - \mathrm{IMI}\left(I\right)\right) \times \sum_{b \in \mathrm{CAB}(I)} P_{\mathrm{eAd}}\left(b,I\right) \qquad (4.10)$$

$$P_{\mathrm{IO}}\left(I\right) = \left(1 - \mathrm{IMI}\left(I\right)\right) - \left(P_{\mathrm{AF}}\left(I\right) + \left(\mathrm{IMI}\left(I\right) \times P_{\mathrm{AF}}\left(I\right)\right)\right) \qquad (4.11)$$

EPI Results for Different Applications: Figure 4.13 shows that in ADPCM and DCT, several instructions have zero value for EPI. This denotes that *either* these instructions do not have any dependent instructions *or* the dependent instructions mask the errors completely. Such masking dependent instructions can be identified by comparing the plots of EPI in Fig. 4.13 and IMI in Fig. 4.11 corresponding to the same instruction address. These instructions are relatively less important for protection compared to those instructions which exhibit high EPI value. In ADPCM and SAD applications, the EPI value is much lower compared to the EPI plot of the DCT application. In case of the DCT application, the EPI value for many instructions is low and

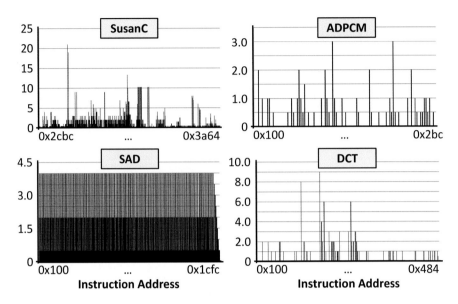

Fig. 4.13 Distribution of the instruction error propagation index (EPI)

the number decreases exponentially as the value grows the same way. This is due to the butterfly form of the instruction dependencies in the DCT data flow graph. This illustrates that it is important to protect earlier executing instructions compared to the later ones. In case of the SAD application, the EPI value is low and the distribution is homogeneous. This is due to the fact that there are many parallel instruction paths with similar dependency structure in the data flow graph. Note that in case of the SAD application, the loops are completely unrolled; therefore, the EPI plot is very dense. The SusanC application exhibits few instructions with high error propagation index having varying distribution and at the same time high IMI and high IVI.

Summary and Observations from the Above Results: The above-discussed IVI, IMI, and EPI models capture the probability of an instruction getting erroneous, error masking and error propagation properties in a given application software program. For selective instruction protection (see Chap. 5), all three models can be jointly taken into account to prioritize instructions for constrained reliability optimization. In case of the ADPCM and SAD applications, the instructions selected for protection can be dominated by the value of IVI, whereas in case of the SusanC application, the EPI values may determine a more feasible instruction selection for protection.

4.4 Function-/Task-Level Reliability Estimation Models

The above-discussed instruction-level models are beneficial for fine-grained reliability optimizing techniques that operate at the instruction granularity, e.g., techniques for reliability-driven compilation. However, in order to enable coarse-grained reliability optimizing techniques that operate at the function/task granularity, e.g., techniques for reliability-driven system software, there is a need for function-level reliability models. Towards this end, this manuscript introduces two reliability models that quantify the reliability at the function level, namely *Function Vulnerability Index* and the *Function Resilience* (see Appendix B).

4.4.1   Function Vulnerability Index

As discussed in Sect. 3.2, faults occurring during the execution of *critical instructions* typically lead to Application Failures, which are more severe compared to Incorrect Output (from the user perspective). Therefore, the *Function Vulnerability Index* (FVI) is computed as the weighted average of function's vulnerability to Application Failures ($\text{FVI}_{\text{Failures}}$) due to *critical instructions* and function's vulnerability to Incorrect Output ($\text{FVI}_{\text{IncorrectOP}}$) due to data errors in the *critical* and *noncritical Instructions*.

$$\text{FVI}_{\text{Failures}} = \sum_{i \in \text{CI}} \text{IVI}_i * P_{\text{Failures}} \left(\text{FaultRate} \right)$$

$$\text{FVI}_{\text{IncorrectOP}} = \sum_{i \in \text{nCI}} \text{IVI}_i * P_{\text{IncorrectOP}-\text{nCI}} \left(\text{FaultRate} \right) \qquad (4.12)$$
$$+ \sum_{i \in \text{CI}} \text{IVI}_i * P_{\text{IncorrectOP}-\text{CI}} \left(\text{FaultRate} \right)$$

$$\text{FVI} = \frac{w_1 \times \text{FVI}_{\text{Failures}} + w_2 \times \text{FVI}_{\text{IncorrOP}}}{\sum I_F} \qquad (4.13)$$

where $\sum I_F$ is the total number of instructions in a function F. The parameters CI and nCI represent the *critical* and *noncritical instructions*, respectively. $P_{\text{Failures}}()$, $P_{\text{IncorrectOP-CI}}()$, and $P_{\text{IncorrectOP-nCI}}()$ represent the probabilities for Application Failure, Incorrect Output due to *critical instructions*, and Incorrect Output due to *noncritical instructions*, respectively, at a given fault rate. These probabilities are obtained through fault injection. Note, the parameters w_1 and w_2 are provided for the designers to prioritize the severity of different error types. This manuscript considers the values of w_1 and w_2 equal to "1" to avoid any kind of weighting effects, i.e., vulnerabilities for both Application Failures and Incorrect Output are considered equally important. The FVI model is used to quantify the effects of several reliability optimizations in this manuscript. Reduced FVI hints towards the effectiveness of the applied reliability optimization technique that in turn reduces the number of Application Failures and/or Incorrect Output, thus achieving an improved software program's reliability.

So far all the above-mentioned reliability metrics quantified the reliability in terms of functional correctness. However, in safety-critical systems, a correct functional output delivered after the deadline represents a degraded time-wise reliability and may not be acceptable or could even be catastrophic. In order to improve the overall reliability, it is important that the correct functional output be delivered within the deadline. Towards this end, the following *Reliability-Timing Penalty* model is devised.

4.4.2 Reliability-Timing Penalty

This model is beneficial for reliability optimizing techniques that target improving the functional reliability while meeting the timing constraints and can explore the tradeoff between the deadline misses versus the functional reliability. The *Reliability-Timing Penalty* (RTP) is the linear combination of the *functional reliability* (i.e., the reliability penalties which could be given as FVI, function resilience, or any other function-level reliability metric), and the *timing reliability* (i.e., the deadline miss probability). Specifically, for a user-defined parameter $0 \le \alpha \le 1$, the RTP for a given function is defined using Eq. 4.14.

$$\text{RTP} = \alpha \text{R} + (1 - \alpha) \text{miss rate}$$

(4.14)

where *miss rate* represents the percentage of deadline misses for the application and R is the function reliability. In case α is closer to 0, the timing satisfaction becomes more important; however, when α is closer to 1, the functional reliability in the presence of faults in the underlying hardware becomes more important. This metric is beneficial in scenarios where the objective of a reliability optimizing technique is to improve the reliability while meeting the timing constraints, i.e., meeting the deadlines. That is, targeting the minimization of the RTP.

4.5 Chapter Summary

This chapter presented different instruction-level reliability models like *Instruction Vulnerability Index*, *Instruction Error Masking Index*, and *Instruction Error Propagation Index* that quantify the error probability, error masking probability, and error propagation effects at the instruction-level granularity. Unlike the proposed models, state-of-the-art does not facilitate comprehensive quantification of these three key reliability properties at the application program granularity. Moreover, state-of-the-art program-level reliability models do not jointly account for the hardware-level knowledge (i.e., where faults occur) and the software-level knowledge (i.e., where errors are observed) in a holistic way. In contrast, the proposed *Instruction Vulnerability Index* model quantifies error probabilities of instructions when the software program is executed in a processor pipeline stage by stage, while considering spatial and temporal effects from both hardware and software perspectives. State-of-the-art program-level reliability techniques do not provide this level of detail and therefore cannot optimize for it. The contributions in this chapter further advance the state-of-the-art by statically quantifying the error masking and propagation effects through a comprehensive analysis of data and control flow graph.

Now the question arises, whether and how these models can be leveraged for software program reliability optimization. The answer to this question is motivated by a detailed experimental analysis of these three reliability models for different application programs. This analysis illustrates that, besides across different applications, even different instructions inside the same application exhibit varying vulnerability, error masking, and error propagation properties due to their distinct utilization of pipeline components, dependent instructions, and control flow. This knowledge can be exploited to quantify the reliability-wise importance of different instructions, which opens a pool of opportunities to devise and employ instruction-level reliability optimization techniques that has not been explored earlier. For instance, transformations and instruction scheduling inside a compiler can be redesigned towards reliability optimization by modifying the instruction profile considering the spatial/temporal vulnerabilities of instructions and reducing the number of critical instruction executions (as will be shown in Chap. 5).

Afterwards, a function-/task-level reliability model *Function Vulnerability Index* is introduced which jointly accounts for the vulnerabilities of all instructions in the given function along with the probability of Application Failures and Incorrect Outputs. In order to incorporate the deadline effects in timing-critical systems, the *Reliability-Timing Penalty* model is introduced that jointly considers the functional and timing reliability (i.e., probability of deadline misses). Note, joint consideration of functional and timing reliability has not been explored by existing state-of-the-art techniques.

The above-discussed models enable prioritization of different instructions and functions/tasks and will be leveraged in Chaps. 5 and 6 to design and evaluate reliability optimization techniques at different software layers of the system design abstraction under tolerable performance overhead constraints.

Chapter 5
Software Program-Level Reliability Optimization for Dependable Code Generation

State-of-the-art has primarily exploited the compiler-level techniques for improving the performance and energy. This chapter aims at enabling reliability-driven compilation enabled by the instruction-level reliability models of Chap. 4 that quantify the reliability-wise importance of different instructions and the impact of their interdependencies on the vulnerability variations. This chapter presents several novel techniques for *reliable code generation* in order to increase software program's reliability under user-provided tolerable performance overhead constraints. Improved software reliability can be achieved in two orthogonal and equally important ways: (1) Reducing the error probabilities by reducing the vulnerabilities to soft errors and *critical instruction* executions; and (2) Error detection and recovery through instruction duplication or triplication, where selective redundancy can be applied to reduce the performance overhead.

In order to reduce the error probabilities, first, four different reliability-driven transformations (Sect. 5.1) are employed that reduce the spatial and temporal vulnerabilities of instructions in different pipeline stages along with reducing the *critical* instruction executions. These four transformations are: (1) *Reliability-Driven Data Types Optimization* that transforms data types with smaller bit-widths into data types with larger bit-widths in order to curtail the number of critical instruction executions. (2) *Reliability-Driven Loop Unrolling* that determines an appropriate unrolling factor such that the total function vulnerability is minimized while considering the impact of unrolling on the spatial and temporal vulnerabilities w.r.t. different processor components, for instance, more/less *live registers* with more/less *vulnerable lifetime*. (3) *Reliability-Driven Common Expression Elimination and Operation Merging* that, unlike traditional common expression elimination, reduces the vulnerabilities of large-sized sub-expressions and analyzes the impact of recomputation vs. reusing the result of complex common sub-expressions on the spatial and temporal vulnerabilities. (4) *Reliability-Driven Online Table Value Computation* that optimizes the codes with heavy usage of static arrays (i.e., the so-called tables stored statically) w.r.t. the memory vulnerability. State-of-the-art has not yet explored the employment of reliability-driven transformations inside the compilers due to the unavailability of precise instruction-level reliability quantification models and the notion of spatial/temporal vulnerability w.r.t. different processor components. This manuscript aims at making contributions towards this end. The key challenge is to determine how appropriate transformations should be applied (e.g., what is a reliability-wise good unrolling factor) such that spatial and temporal vulnerabilities for all the instructions inside a given function are collectively minimized. Therefore, a comprehensive vulnerability analysis for different transformations is also required, which is performed in the corresponding sections to understand the reliability-performance tradeoffs of these transformations.

Since an instruction schedule determines the sequence of instructions to be executed while considering their interdependencies, this can potentially affect the vulnerabilities of different instructions (in particular temporal vulnerabilities). However, reducing the vulnerability of an instruction may lead to an increase in the vulnerability of another instruction. Therefore, a reliability-driven instruction scheduler needs to account for such factors from a global perspective, rather than a local one. Existing instruction scheduling techniques for reliability improvement only reorder the code after performance-driven scheduling, which due to the reduced slack (already optimized in the first scheduling phase) have limited potential (typically 2–7 %). These techniques lack in-depth knowledge of vulnerability interdependencies of different independent and dependent instructions. Towards this end, a lookahead-based reliability-driven scheduler (Sect. 5.2) is proposed in this chapter that prioritizes instructions for scheduling based on their vulnerabilities, their categorization w.r.t. the knowledge of critical and non-critical instructions, and the number of dependent instructions with their corresponding reliability profits. To further improve the reliability, error detection and recovery features are provided through selective instruction redundancy under performance overhead constraints. Unlike state-of-the-art techniques that employ full-scale instruction redundancy, the proposed reliability models in this manuscript enable characterizing different instructions based on their vulnerability, error masking, and propagation properties. This enables selecting the reliability-wise most important instructions first, thereby leading to high reliability efficiency in constrained scenarios. A selective instruction redundancy technique is presented in Sect. 5.3 that prioritizes different instructions in a data flow graph w.r.t. their reliability importance (characterized as a joint function of instruction vulnerabilities, error masking, and propagation indexes) under a performance overhead constraint.

These reliability-enhancing techniques are applied to generate a set of multiple compiled versions for each application function (Sect. 5.4), such that, different functions exhibit distinct reliability and execution time properties. This provides a reliability-performance optimization space, which is later exploited by the system software layers to achieve high reliability efficiency during application execution under constrained scenarios.

In the following subsections, these different software-level reliability optimizing techniques are explained in detail.

5.1 Reliability-Driven Software Transformation

The traditional software transformation techniques mainly perform code optimization from the perspective of high performance and low power consumption. This manuscript aims at *rethinking* the traditional software transformation techniques from the reliability perspective, i.e., to increase the reliability of fault-susceptible hardware/software systems through modifying the instruction profile. For reliability quantification, these techniques employ the proposed software reliability metrics of Chap. 4, i.e., *Instruction Vulnerability Index* (*IVI*) and *Function Vulnerability Index* (*FVI*) that quantify the reliability at instruction and function granularity while jointly accounting for software program-level and hardware-level knowledge.

The application error analysis mentioned in Sect. 3.2 illustrates that an increased number of *critical instruction* executions and high spatial/temporal vulnerabilities during the instruction executions lead to increased applications' susceptibility towards Application Failures. The reliability-driven software transformations modify the instruction profile and aim at reducing the number of *critical instruction* executions and instruction vulnerabilities in order to minimize the application's susceptibility towards Application Failures under a tolerable performance overhead constraint. The following four reliability-driven software (source-level) transformations are proposed that can be applied in the front-/middle-end code optimization stage of a compiler to generate reliable assembly code under a given set of constraints.

1. **Reliability-Driven Data Type Optimization** transforms the data types for a given data structure that ultimately impacts the amount of data to be loaded from and/or stored into the memory along with the number of instruction executions to use this data. It reduces the number of *critical instruction* executions while minimizing the function vulnerability (Sect. 5.1.1).
2. **Reliability-Driven Loop Unrolling** determines an "appropriate" unrolling factor such that the function vulnerability is minimized while reducing the number of *critical instruction* executions. It aims at reliability optimization under tolerable performance overhead and code size constraints (Sect. 5.1.2).
3. **Reliability-Driven Common Expression Elimination and Operation Merging** determine the decision for each occurrence of a common expression whether it will be eliminated and its result will be reused, or it will be recomputed. The goal is to minimize function vulnerability under constraints of tolerable performance overhead and code size increase (Sect. 5.1.3).
4. **Reliability-Driven Online Table Value Computation** determines whether using values from the precomputed table or online table value computation will result in a reduced function vulnerability (Sect. 5.1.4).

These reliability-driven software transformations reduce both Incorrect Output and Application Failures by minimizing the spatial and temporal vulnerabilities for all types of instructions. For a given application software, in particular, these transformations minimize the FVI through the following means: (1) *Lowering the* P_{Failures} *and* $P_{\text{IncorrectOP-CI}}$ *probabilities* which is achieved by reducing the number of *critical instruction* executions. (2) *Lowering the* $P_{IncorrectOP-nCI}$ *and* IVI_i is achieved by changing the instruction profile which leads to a different usage pattern of the processor components by means of executing alternative instructions. A detailed vulnerability analysis is performed for all of the above-mentioned transformations to illustrate their impact on the vulnerability w.r.t. different processor components, different design points, and tradeoff between performance and reliability. Afterwards, detailed fault injection experiments and error distribution results are presented for various reliable code versions.

5.1.1   Reliability-Driven Data Type Optimization

Definition *Data type optimization is a technique which transforms the data types with smaller bit widths (like 8-bit unsigned char) into the data types with larger bit widths (16-bit unsigned short or 32-bit unsigned int) for a given data structure.* It affects the amount of data to be loaded from and/or stored to the memory that impacts the instructions executed, thus resulting in a different instruction histogram. *Consequently, this leads to a reduced number of critical instruction executions, while minimizing the FVI.*

Example Figure 5.1 presents an example with original and transformed codes with their data flow graphs. The original code executes 2× more load/store instructions because of the 16-bit data loading into one 32-bit register at a time. However, in the transformed code, two 16-bit data values are loaded into a 32-bit register in a packed format. The reduced number of load/store and their dependent ALU instruction executions results in a reduced probability of Application Failures. There can also be cases where error masking occurs: in line 5 of the transformed code, the lower 16 bits of a 32-bit data load result *r0* are extracted. If a fault causes single/multiple bit flips in the higher 16 bits of *r0* while the data resides in the input register of the ALU, it is masked by the AND operation. Likewise, if the second operand is corrupted (i.e., 0x0000FFFF), the fault is masked in case the corresponding value in *r0* is "0".

This transformation brings certain side-effects, i.e., the introduction of additional extraction and merging code (shown in Fig. 5.1b). Later when the data is unpacked, the variables and instructions are still in 32-bit format, the overflow of signed values is avoided. The execution of the additional instructions for packing and unpacking of data incur a performance penalty in addition to a relatively higher vulnerability w.r.t. ALU. Therefore, the associated *overhead due to additional instruction executions needs to be amortized by the vulnerability reduction due to reduced number of executions of load and store instructions.* Load instructions may incur stalls inside the pipeline due to cache misses. Hence, a reduced number of load instruction executions may even amortize the performance overhead associated with the execution of the additional extraction and merging code. This transformation is relatively more beneficial for the VLIW architecture due to the availability of SIMD instructions.

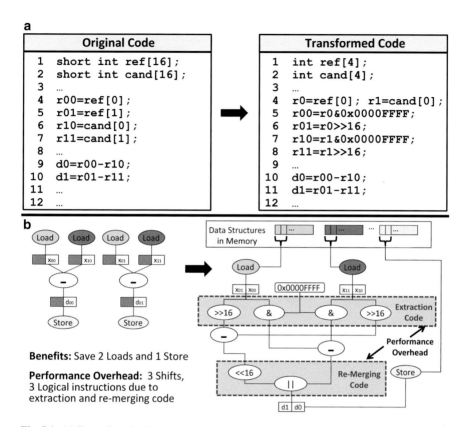

Fig. 5.1 (**a**) Example code showing data type optimization transformation, (**b**) corresponding data flow graphs

Applying FVI-Driven Data Type Optimization: Figure 5.2 presents the flow for applying the optimization targeting load merging; for store instructions, the procedure is similar (see detailed algorithm in Appendix C). It operates on a function graph $G(V, E)$ given a tolerable performance overhead P_τ, FVI and performance of the original code (FVI_{Orig}, P_{Orig}). The output of this optimization is a transformed function fd with merged loads and extraction code as a result of the data type optimization. First, all arrays are extracted from the function graph. For each array a, a list L of all load vertices is obtained. For evaluation, a temporary copy G' of the graph is created. Then, two consecutive load vertices are extracted from the load list, merged, and inserted into the temporary graph G' along with the extraction code. Afterwards, the function is compiled and simulated to estimate the performance and reliability. If the given constraints are satisfied and a better solution is found, the vertices under evaluation are removed from the original graph G and the merged load vertex is inserted along with the extraction code.

Vulnerability Analysis of the Transformation: Figure 5.3 shows the average IVI comparison for the *untransformed code* (i.e., with character data type) and the

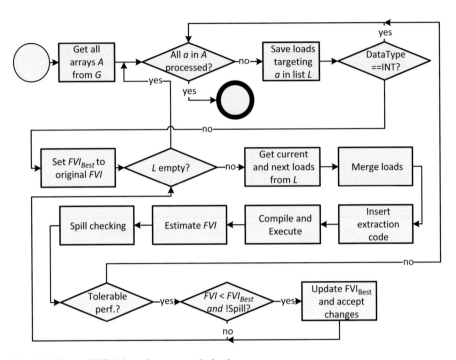

Fig. 5.2 Flow of FVI-driven data type optimization

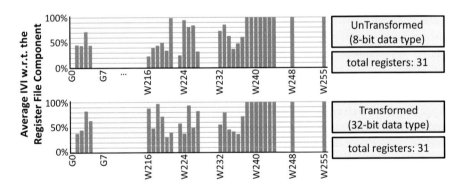

Fig. 5.3 IVI w.r.t. different registers: character (8-bit) vs. integer (32-bit) data type

transformed code (i.e., with integer data type) w.r.t. the register file component. In both cases, 31 registers are used for storing the variables. However, the IVI w.r.t. the register file for the transformed code is slightly higher in contrast to that of the untransformed code (i.e., 8.20 % vs. 7.92 %). The total vulnerable period w.r.t. the register file component becomes higher, i.e., from 4752 cycles to 4963 cycles (i.e., a 4 % increase) for a single function execution. This is because, in the transformed code, 4 data bytes are loaded into a 32-bit register using only one instruction and the data resides longer in the register until all the data-dependent instructions finish their executions and the register is overwritten. Moreover, the IVI in the execute stage is higher because of the additional shift and logic instructions required for the "data unpacking." However, this transformation significantly reduces the IVI w.r.t. the cache controller and the number of Application Failures.

5.1.2 Reliability-Driven Loop Unrolling

Definition *It is a technique to unroll the source code loops by an appropriate unrolling factor amongst various unrolling options, such that, the Function Vulnerability Index (FVI) is minimized, while reducing the number of critical instruction executions.* The unrolling factor is the number of loop body replications once it is unrolled. Traditionally, loop unrolling has only been done to improve the performance; however, the problem of finding an appropriate unrolling factor has not yet been explored from the reliability perspective.

Example On the one hand, loop unrolling reduces the number of *critical instruction* executions, loop counter update, and the condition evaluation operations that lower the error probability and execution probability of backward jumps to the loop condition. On the other hand, it may result in an increased FVI due to the following two reasons: (1) Increased *temporal vulnerability* of variables stored in the register file as they are typically kept for a relatively longer time inside the registers until the relevant instructions are executed. The example code in Fig. 5.4 shows an increase in the vulnerable period of variable $y[2]$. In contrast, the original code reloads $y[2]$ while exhibiting a reduced vulnerable period. (2) Increased *spatial vulnerability* as more registers may be required for storing live variables. The example code shows that more $y[i]$ data values are alive, thus requiring more registers.

FVI-Driven Loop Unroller: *The challenge in this case is to determine an appropriate **unrolling factor** which is driven by the FVI such that the reliability is optimized while taking into account the relative increase in execution time and code size.* In order to address the above-discussed reliability-related concerns of loop

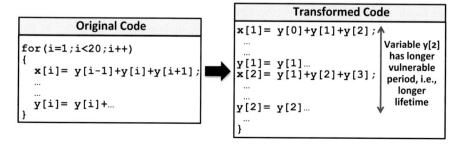

Fig. 5.4 Example: an abstract code showing increased temporal and spatial vulnerabilities of variables as a consequence of loop unrolling

unrolling, an *FVI-driven Loop Unroller* is proposed that determines an appropriate unrolling factor for each given loop of a function while minimizing the following two factors: (1) FVI considering utilization of different processor components by different instructions; and (2) Performance loss compared to the maximum achievable performance when using a performance-based unrolling. It aims at avoiding spilling and incurring a relatively small increase in code size (i.e., number of assembler instructions). The *FVI-Driven Loop Unroller* discards the unrolling factors that cause register spilling, as it may result in the execution of additional *critical instructions* such as *store* and then *load*. Thus, the probability for Application Failures might increase because of the spill code. The objective is to maximize the following *profit* function (Eq. 5.1). {FVI_{Orig}, P_{Orig}, C_{Orig}} and {FVI, P, C} represent the FVI, performance, and code size of the original code and the transformed code, respectively. The parameter μ *activates* the normalization effect because of the code size expansion in case the instruction cache is not protected, or *deactivates* in case the instruction cache is protected through ECC or parity-based techniques.

$$\Pr ofit = \left(\frac{FVI_{Orig} - FVI}{FVI_{Orig}} \right) \bigg/ \left(\mu \times \frac{C - C_{Orig}}{C_{Orig}} + \left(\frac{P - P_{Orig}}{P_{Orig}} \right) \right) \qquad (5.1)$$

Applying FVI-Driven Loop Unrolling (Fig. 5.5): The proposed *FVI-driven Loop Unroller* requires the *loop iteration counts*. This is known for fixed-sized and input-invariant loops and *unknown* for variable-sized loops where the loop iterations depend upon the input data.[1] For each loop of a given function, the maximum unrolling factor is then determined as the *Greatest Common Divisor* of all the corresponding loop iterations obtained through profiling for varying input data. The inputs are a set of maximum unrolling factors for all loops, the FVI, performance, and code size of the original function *F*. The output is the transformed function *fd* with loop unrolling applied by an *FVI-minimizing unrolling factor*. First, all loops are extracted and stored in a list *L*. Afterwards, all loops of the function are processed and an appropriate unrolling factor is determined.

For each loop, the corresponding maximum unrolling factor is extracted from the input set of maximum unrolling factors. Then, for each loop, the profit for all possible values of the unrolling factors is computed. For each case of the unrolling factor, the FVI reduction, performance loss, and code expansion are estimated to compute the profit value and the one with the highest profit is selected as the *FVI-driven best unrolling factor*. Afterwards, the next loop is analyzed. Note that the proposed approach discards the spilling cases. The detailed algorithm is provided in Appendix C.

Vulnerability Analysis of the Transformation: Figure 5.6 shows the comparison for the average IVI w.r.t. the register file component for the untransformed code (i.e., with rolled loop) and the transformed code (i.e., with loop unrolling by factors 4 and 8). In

---

[1] Identifying fixed loops and variable loops is out of the scope of this work; see [78, 92] for further details on such a static loop analysis.

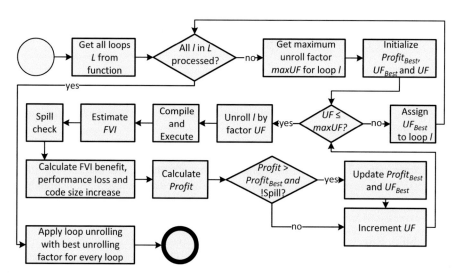

Fig. 5.5 Flow of FVI-driven loop unrolling

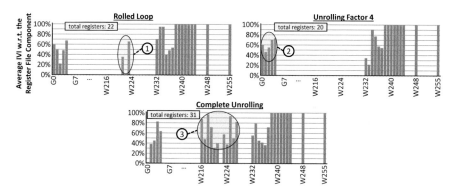

Fig. 5.6 Comparing the average IVI w.r.t. the register file component for the untransformed code (loop rolled) vs. transformed code (loop unrolling by factors 4 and 8) for the "SATD" application

case the loop is unrolled by a factor 4, the vulnerability of registers $G1$–$G4$ is increased. This is because these registers are used for a longer time, storing the temporary values (see label ②). In addition to this, two less registers are used in the untransformed code, i.e., 20 vs. 22, and this leads to a slightly reduced spatial vulnerability. It is further observed that, in the untransformed code (see Label ①) there is a higher IVI value for the registers $W221$ and $W223$. These registers are used to store the loop counters and memory addresses. Unrolling the loop by a factor of 4 results in a reduction from 6 to 5.63 % for the IVI w.r.t. the register file. However, increasing the unrolling factor from 4 to 8 also increases the register usage, i.e., from 20 to 31, which leads to a higher spatial vulnerability (see label ③). Consequently, the total IVI of the complete unrolling case becomes worse when compared to the unrolling by a factor of 4, i.e., 8.20 % vs. 5.63 %. In this example, 4 is the reliability-wise best unrolling factor.

5.1.3   Reliability-Driven Common Expression Elimination and Operation Merging

Definition *Operation Merging and Common Expression Elimination is a method to identify and eradicate identical expressions with multiple operations and to merge the execution of two identical operations into one such that the total function vulnerability is minimized.*

Example 1 Figure 5.7 presents an example for the matrix multiplication in "Hadamard Transformation." The matrix multiplications are expanded and transformed into expressions where only *add* and *shift* instructions are used and *multiply* instructions are completely replaced. Afterwards, the expressions are reorganized in a way that common expressions are identified and eliminated, and the common operations are merged.

Reliability Effects of the Transformation: This transformation can affect the spatial and temporal vulnerabilities in different ways. On the positive side, it reduces the FVI via the following three means: (1) reducing excessive computation of *arithmetic/ALU* instructions that lead to a reduced number of Incorrect Outputs or Silent Data Corruptions, therefore, reducing the vulnerability w.r.t. ALU; (2) reducing *multiply* instructions, that corresponds to the reduction of spatial and temporal vulnerabilities in the execute stage; and (3) minimizing the number of *load* instruction executions due to fewer accesses to the input variables stored inside the memory that were previously used to compute the common expressions, thus reducing $P_{Failures}$. However, on the negative side, due to the excessive usage of temporary register variables holding the common values for reusability, the spatial vulnerability w.r.t. the register file is increased. Furthermore, it may lead to an increased temporal vulnerability if the register storing the result of the common expression is used much later in the instruction schedule. A high register pressure may even lead to register spilling resulting in an increased number of *critical instruction* executions. In such cases, it is beneficial to re-compute the common value.

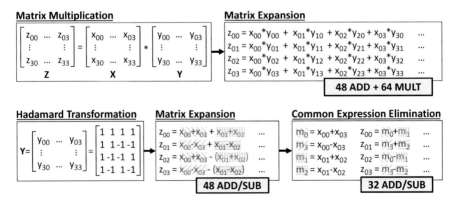

Fig. 5.7 Common expression elimination in the Hadamard transformation

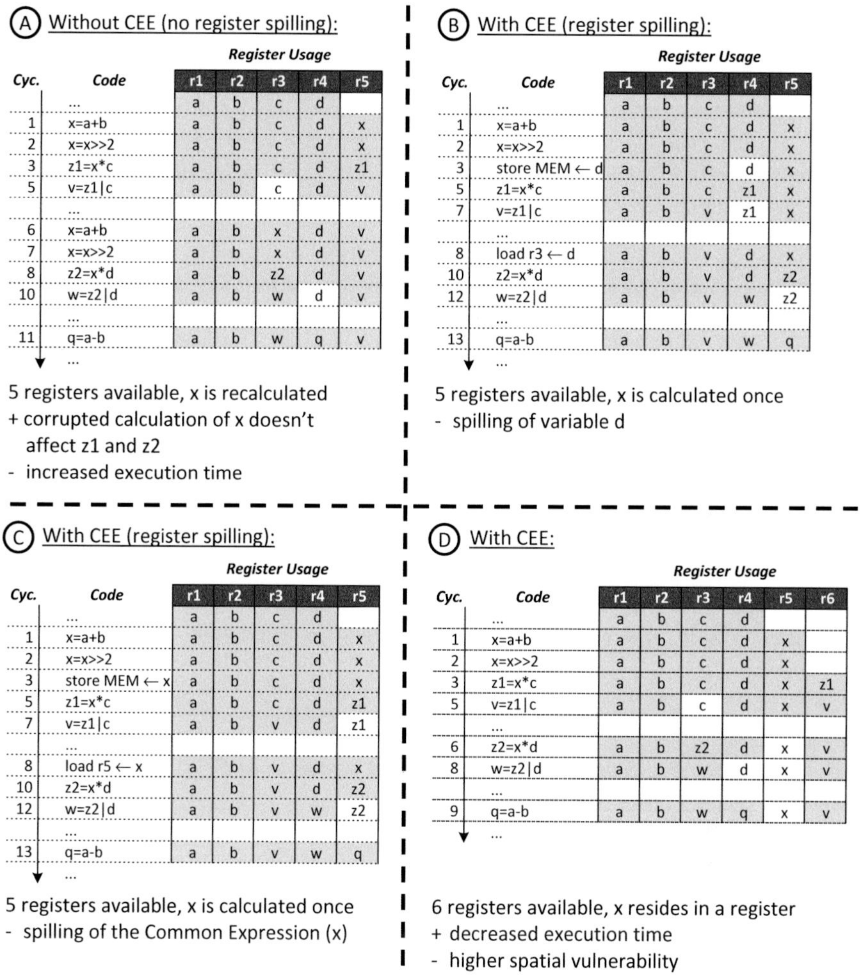

(A) Without CEE (no register spilling):

		Register Usage				
Cyc.	Code	r1	r2	r3	r4	r5
...		a	b	c	d	
1	x=a+b	a	b	c	d	x
2	x=x>>2	a	b	c	d	x
3	z1=x*c	a	b	c	d	z1
5	v=z1|c	a	b	c	d	v
6	x=a+b	a	b	x	d	v
7	x=x>>2	a	b	x	d	v
8	z2=x*d	a	b	z2	d	v
10	w=z2|d	a	b	w	d	v
11	q=a-b	a	b	w	q	v
...						

5 registers available, x is recalculated
+ corrupted calculation of x doesn't
  affect z1 and z2
- increased execution time

(B) With CEE (register spilling):

		Register Usage				
Cyc.	Code	r1	r2	r3	r4	r5
...		a	b	c	d	
1	x=a+b	a	b	c	d	x
2	x=x>>2	a	b	c	d	x
3	store MEM ← d	a	b	c	d	x
5	z1=x*c	a	b	c	z1	x
7	v=z1|c	a	b	v	z1	x
8	load r3 ← d	a	b	v	d	x
10	z2=x*d	a	b	v	d	z2
12	w=z2|d	a	b	v	w	z2
13	q=a-b	a	b	v	w	q
...						

5 registers available, x is calculated once
- spilling of variable d

(C) With CEE (register spilling):

		Register Usage				
Cyc.	Code	r1	r2	r3	r4	r5
...		a	b	c	d	
1	x=a+b	a	b	c	d	x
2	x=x>>2	a	b	c	d	x
3	store MEM ← x	a	b	c	d	x
5	z1=x*c	a	b	c	d	z1
7	v=z1|c	a	b	v	d	z1
8	load r5 ← x	a	b	v	d	x
10	z2=x*d	a	b	v	d	z2
12	w=z2|d	a	b	v	w	z2
13	q=a-b	a	b	v	w	q
...						

5 registers available, x is calculated once
- spilling of the Common Expression (x)

(D) With CEE:

		Register Usage					
Cyc.	Code	r1	r2	r3	r4	r5	r6
...		a	b	c	d		
1	x=a+b	a	b	c	d	x	
2	x=x>>2	a	b	c	d	x	
3	z1=x*c	a	b	c	d	x	z1
5	v=z1|c	a	b	c	d	x	v
6	z2=x*d	a	b	z2	d	x	v
8	w=z2|d	a	b	w	d	x	v
9	q=a-b	a	b	w	q	x	v
...							

6 registers available, x resides in a register
+ decreased execution time
- higher spatial vulnerability

Fig. 5.8 Impacts of different strategies for exploitation of common expressions

Example 2 Figure 5.8 presents different scenarios showing the possible impacts of common expression elimination (CEE). The functionality of the example codes is the same for all cases. For the cases (A)–(C) five registers can be used, while in case (D) six registers are available. In example (A), the option to remove the common expression is not exploited. Therefore, a potentially corrupted computation of the variable x does not affect several data-dependent computations. In examples (B) and (C), the common expression is eliminated; however, this results in register spilling. For example, in (B) the register storing the variable d is spilled that results in a longer execution time. However, in example (C) the register holding the result of the common expression (i.e., variable x) is spilled. Consequently, this leads to a high potential of affecting a large amount of dependent calculations in case the storing/loading process is corrupted.

The register spilling can be avoided by using one additional register which will result in a shorter execution time; however, this will lead to an increased spatial vulnerability. For reliability optimization, the complier needs to be judicious about whether to perform the common expression elimination or not. *The challenge, in this case, is to decide for each occurrence of a common expression whether it will be eliminated and its result will be reused, or it will be recomputed. The goal is to minimize the FVI under constraints of tolerable performance overhead and increase in the code size.*

Applying FVI-Driven Common Expression Elimination: Figure 5.9 shows the flow for evaluating the reliability benefit of replacing common expressions in a given function graph G under a tolerable performance overhead P_τ (see detailed algorithm in Appendix C). The process operates in two phases: An *evaluation phase* and an *elimination phase*. In the evaluation phase, all common expressions and their respective occurrences are extracted. Afterwards, the potential for replacing each individual occurrence of a common expression is evaluated by comparing the function vulnerability with and without replacement. Before the elimination phase, the common expressions extracted are sorted by their reliability efficiency in descending order. If a given solution of the common expression elimination satisfies the constraints and register spilling is avoided, it is accepted and the common variable is inserted. Finally, the resulting code with expression elimination is returned.

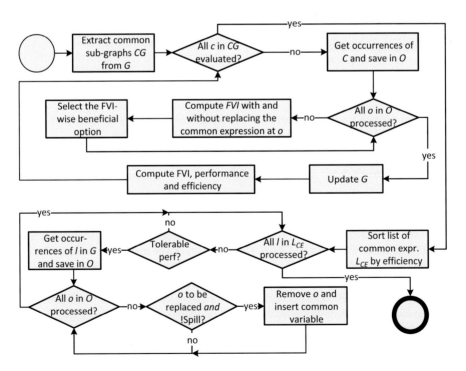

Fig. 5.9 Flow of applying common expression elimination

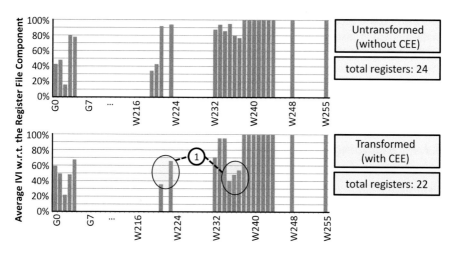

Fig. 5.10 Instruction histogram and IVI of each register: common expressions elimination and operation merging

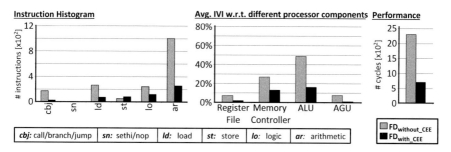

Fig. 5.11 Effect of common expression elimination and operation merging on reliability and performance

Vulnerability Analysis of the Transformation: Figure 5.10 presents the average IVI comparison w.r.t. the register file component *with* and *without* the common expressions, respectively. As an example, we have analyzed the *SATD* application. This transformation replaces the common expressions associated to the redundant *multiply* instructions with *add* and *shift* instructions. Consequently, the number of *multiply* instructions are reduced that has a high spatial and temporal vulnerability due to (a) its multi-cycle nature and (b) large chip area footprint of the multiply unit. It results in the vulnerability reduction of registers W221, W224, W235, W236, and W237 from 87 to 48 % (see label ①). This transformation reduces the spatial vulnerability by using two less registers. The total IVI w.r.t. the register file is reduced, i.e., from 7.24 to 6.05 %.

Figure 5.11 presents the instruction histogram, IVI for different processor components, and performance. This transformation reduces the number of *arithmetic/logic* instruction executions. Furthermore, the *load* instructions can also be reduced because of fewer accesses to the input variables stored inside the memory. A reduction in the amount of instruction executions will ultimately lead to reduced temporal vulnerability but at the cost of a higher potential for data error propagation to multiple dependent instructions, in case a common variable is corrupted. Nevertheless, the IVIs with respect to all processor components are reduced.

5.1.4   Reliability-Driven Online Table Value Computation

Definition *Online table value computation is a method that calculates values in the table entries indexed in a program during the execution, rather than loading the values from precomputed table entries.*

Reliability Effects of Precomputed Tables: In contrast to the online table value computation, the positive aspect of this method is that it offers better performance. Moreover, due to less register usages the benefit can come in two ways: (1) reduced IVI w.r.t. the register file, (2) reduced risk of register spilling, as one register is used for storing the base address. However, on the negative side, more *load* instructions are executed that may lead to a high P_{Failures}. Moreover, in case the cache or on-chip memory is not protected, the table will be vulnerable for a long time. This leads to increased spatial and temporal vulnerability of instructions in the memory stage. Due to similar reasons, in case of security-based applications, it may be beneficial to perform online table value computation.

Reliability Effects of Online Table Value Computation: On the positive side, this method requires lesser memory and provides a reduced IVI in the memory stage. Moreover, due to the reduced number of *load* instruction executions, this method reduces the P_{Failures}. However, on the negative side, this method incurs performance overhead because of the additional arithmetic instructions. Consequently, this leads to a higher IVI in the execute stage (w.r.t. the ALU) in contrast to the precomputed table method. However, due to the inherent program-level masking effects, the corrupted result might not propagate and appear at the final program output. Due to high register usage and lifetime, the other side effect of this method is an increased spatial and temporal vulnerability w.r.t. the register file component. Moreover, this may increase the risk of register spilling. In online value computation, the data memory is not used. The only possibility from where the faults could stem is either from instruction memory, where a corrupted instruction results in an Application Failure, or the fault might come during online computation. However, in a precomputed table value, the fault can stem from a corrupted data memory. *The challenge is to determine whether using precomputed table values will minimize the FVI or the online table value computation.* To address this, the FVI values of both versions of the function are computed and the one with the lowest FVI value is selected.

Example and Experimental Analysis: Figure 5.12 presents an example for the *ADPCM* application comparing the precomputed table with the online table value computation. In case of the precomputed table, the total number of executions of load

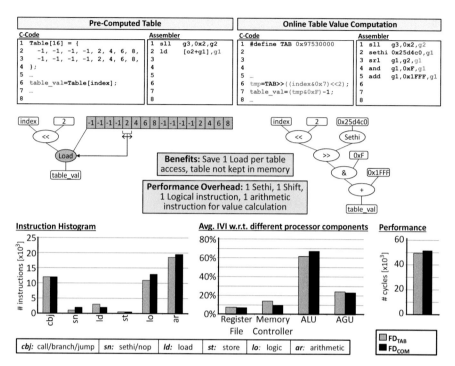

Fig. 5.12 Comparing precomputed table and online table value computation methods for the "ADPCM" application

instructions is higher compared to the other case. Hence, this results in a higher IVI w.r.t. the memory controller. Moreover, the base address along with the index value of the table is used to compute the load address. The base address is kept inside a register throughout the execution of the application. Consequently, the register is kept alive throughout the execution of the application. Hence, a slight increase in the IVI w.r.t. the register file can be observed. However, in the online table value computation, the number of *critical instruction* executions is reduced, as the table entries are computed at run-time. This leads to a reduced IVI w.r.t. the memory controller. However, a slight increase in the IVI w.r.t. the ALU is observed, since more arithmetic operations are performed. The IVI w.r.t. the register file is fairly similar in both cases because the register usage patterns for both function versions of *ADPCM* are similar.

Benefits in case of Unprotected Memories: Figure 5.13a presents a comparison between the precomputed table, i.e., the baseline case (B), and online table value computation (V1) for two applications, i.e., *ADPCM* and *CAVLC* (Context Adaptive Variable Length Coding). The tables are placed in unprotected memory areas. Besides the reduction of Application Failures, the amount of Incorrect Outputs is also decreased once the proposed transformation is applied. Furthermore, Fig. 5.13b shows that in case of the precomputed table in the *ADPCM*, the data remains vulnerable for a long time. The temporal vulnerability of the respective table entries

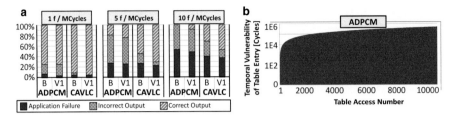

Fig. 5.13 (**a**) Error distribution for "ADPCM" and "CAVLC" when applying FVI-driven online table value computation; (**b**) temporal vulnerability (in cycles) of the entries requested for first 10,000 table accesses

requested for the first 10,000 accesses is recorded. The vulnerable time for the table entries is increasing with longer execution runs and entries used by the later accesses are more susceptible to soft errors. In contrast, when replacing the table by online calculations, the vulnerable time of two registers used $g1$ and $g2$ is only a few cycles, as no constant stays in a register or in the memory for the complete application execution and no base address for the table has to be maintained leading to a reduced temporal vulnerability for the calculated values.

5.1.5   Impact of Reliability-Driven Transformations on Error Distributions

Figure 5.14 presents detailed error distribution for different application versions listed in Table 5.1. To provide a fair comparison, the original code (i.e., the baseline case B) is optimized using basic performance optimizations done by the programmer (like, replacing the multidimensional arrays with single-dimensional arrays to avoid multiple-indirections in the memory access and replacement of integer multiply with shift and add where possible) and the traditional compiler optimizations (like common sub-expression elimination).

A general observation is that for most of the applications, the number of Application Failures can be reduced. However, the application executions resulting in an Incorrect Output are partially increasing in a few cases. The focus of the above-mentioned transformations is to reduce the number of *critical instruction* executions such that the Application Failures are avoided. In case of *ADPCM* and *SAD,* the benefit of this transformation is not as significant as for the other applications. The reason is that in these applications the number of *critical instruction* executions are admittedly reduced; however, more arithmetic/logic instructions are executed resulting in, e.g., a higher potential for nondecodable instructions. For *ADPCM*, the online table computation has a slight reliability improvement for higher fault rate cases, i.e., 5 and 10 f/MCycles. In order to achieve further reliability benefits, other transformations need to be applied in addition to this transformation. Furthermore, for *ADPCM*, V2 is better than V1 and B because of the data type optimization (for 10 f/MCycles, V4 is better than V3). However, when the data loads are changed from 32-bit to

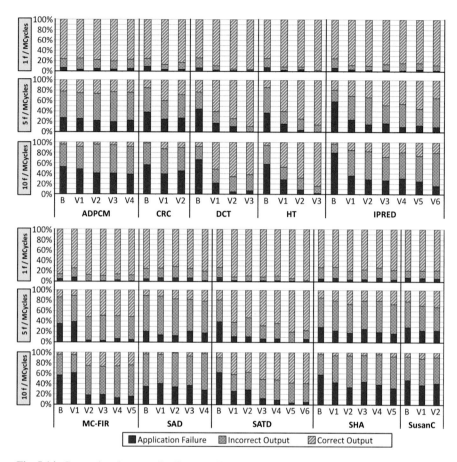

Fig. 5.14 Comparing the error distribution of baseline and the reliability-optimized functions

16-bit, the benefit does not seem significant. Therefore, this exploration shows that the data type optimization using 32-bit data loads is better in this particular scenario. In general, data type optimization is more beneficial in memory intensive applications and even more beneficial in architectures with support for SIMD operations. In case of the *CRC* application, the data type transformation effectively reduces both the Incorrect Outputs and Application Failures. The modified table implementation increases the size of the table and performs online calculations aiming at increasing the performance and reduced number of table accesses. Therefore, it does not provide the desired effects especially related to the Incorrect Output results that can be attributed to the larger table and additional online calculations.

For *DCT* and *HT*, a significant improvement can be observed when applying the transformations in a stepwise fashion. There are two reasons for that: (a) the number of *critical instruction* executions is significantly reduced and, (b) the vulnerable periods are decreased. Same observations hold true for *SATD* and *MC-FIR* applications. In case of *IPRED*, the highest improvement is achieved through stepwise reduction of the

Table 5.1 Overview of the application versions created (the reliability-wise best version is highlighted)

Appl.	Ver.	Description
ALL	B	Baseline
ADPCM	V1	B + online table value computation
	V2	V1 + data type optimization (32 bit loads instead of 16 bit)
	V3	V2 + data type optimization (storing 16 bit data instead of 8 bit), loop unrolling (factor 4)
	V4	*V3 + data type optimization (storing 32 bit data instead of 16 bit)*
CRC	*V1*	*B + data type optimization (32 bit loads instead of 16 bit)*
	V2	V1 + larger lookup table for 16 bit data instead of 8 bit with 8192 entries + online calculation
DCT	V1	B + common expression elimination
	V2	*V1 + loop unrolling (factor 2)*
	V3	V1 + complete loop unrolling (factor 8)
HT	V1	B + common expression elimination
	V2	V1 + loop unrolling (factor 2)
	V3	*V1 + complete loop unrolling (factor 8)*
IPRED	V1	B + loop unrolling (factor 4)
	V2	Data type optimization (32 bit loads instead of 8 bit)
	V3	*B + loop unrolling (factor 16 + 4)*
	V4	V1 + loop unrolling (factor 16)
	V5	V2 + calculation reordering[a]
	V6	*V3 + calculation reordering*
MC-FIR	V1	B + data type optimization (32 bit loads instead of 8 bit)
	V2	B + common expression elimination
	V3	V1 + common expression elimination
	V4	*V2 + loop unrolling (factor 8 + 4)*
	V5	*V3 + loop unrolling (factor 8 + 4)*
SAD	V1	B + data type optimization (32 bit loads instead of 8 bit)
	V2	V1 + loop unrolling (factor 16)
	V3	*B + calculation reordering*
	V4	*V3 + loop unrolling (factor 16)*
SATD	V1	B + common expression elimination, loop unrolling (factor 4)
	V2	V1 + data type optimization (32 bit loads instead of 8 bit)
	V3	V1 + additional loop unrolling (factor 4 + 4)
	V4	V2 + additional loop unrolling (factor 4 + 4)
	V5	V3 + complete loop unrolling (factor 4 + 16)
	V6	*V4 + complete loop unrolling (factor 4 + 16)*
SHA	V1	B + loop unrolling (factor 80)
	V2	V1 + additional loop unrolling (factor 80 + 80)
	V3	V1 + common expression elimination
	V4	V3 + additional common expression elimination
	V5	*V2 + V4*
SusanC	*V1*	*B + common expression elimination, loop unrolling (factor 2)*
	V2	V1 + additional common expression elimination, data type optimization (storing 16 bit data instead of 8 bit)

[a]The "calculation reordering" refers to a modified order in which the different algorithm/calculation steps are executed.

control flow and store instructions for the first two versions. The remaining loop unrolling and computation sequence changes also alleviate the Application Failures. In case of *SHA* the combination of (excessive) loop unrolling and common expression elimination turns out to be a reasonable combination. For *SusanC* the common expression elimination transformation (that aims at decreasing the number of load/store critical instructions) applied in the first step improves the behavior of the application. While the second application version still shows improvements compared to the baseline implementation, the overhead for control flow instructions is significantly increasing. This results in a worse alternative even though the number of *load/store* instructions could be reduced further. Figure 5.15 provides the detailed distribution of different types of Application Failures for *SATD* and *SHA*. A comparison between the baseline and a version with all transformations is shown. In case of *SATD*, it is observed that for almost all types of errors, the Application Failures can be reduced significantly due to the reduction of *load/store* and control flow instructions. However, the remaining Application Failures are primarily due to non-decodable instructions, which cannot be solved with the proposed transformations. Figure 5.14 shows that the proposed transformations offer on average 63.8 %, 81.9 %, and 67.8 % reduced Application Failures for 1, 5, and 10 f/MCycles, respectively. Additionally, these transformations provide on average 43.7 %, 16.2 %, and −38.6 % reduced (negative value denotes an increase) Incorrect Outputs for 1, 5, and 10 f/MCycles, respectively.

Note that the "calculation reordering" applied for *SAD* in V3 (see Table 5.1) refers to a modified order in which the different algorithm/calculation steps are executed. The *SAD* Baseline (B) implementation is compared with the *SAD* V3 implementation in Fig. 5.16. For "*SAD* Baseline," in the first step, the data is extracted from an array (*c00-c03*; *r00-r03*) and stored in variables, which are used

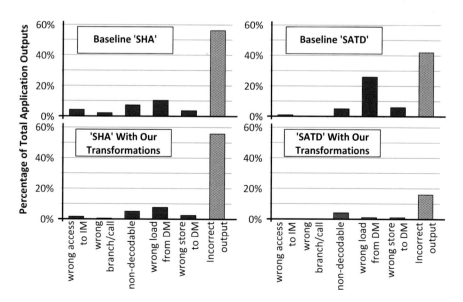

Fig. 5.15 Comparing the distribution of different error types for baseline and the reliability optimizations for SHA and SATD

SAD Baseline		SAD V3	
C-Code	Assembler	C-Code	Assembler
1 char c00, c01, c02, c03;	1 ldub [%i0+(0x1)],%i5	1 c=cMB[0]; r=rMB[0];	1 ldub [%o0+%g0],%g2
2 char r00, r01, r02, r03;	2 ldub [%i0+(0x2)],%i3	2 SAD+=ABS((int)c-(int)r));	2 ldub [%o1+%g0],%g1
3 c00=cMB[0]; c01=cMB[1];	3 ldub [%i0+(0x3)],%o7	3 c=cMB[1]; r=rMB[1];	3 sub %g1,%g2,%g3
4 c02=cMB[2]; c03=cMB[3];	4 ldub [%i1+(0x1)],%g3	4 SAD+=ABS((int)c-(int)r));	4 subcc %g2,%g1,%g2
5 r00=rMB[0]; r01=rMB[1];	5 ldub [%i1+(0x2)],%i4	5 c=cMB[2]; r=rMB[2];	5 bpos,a 0x120
6 r02=rMB[2]; r03=rMB[3];	6 ldub [%i0+%g0],%g2	6 SAD+=ABS((int)c-(int)r));	6 mov %g2,%g3
7 SAD+=ABS((int)c00-(int)r00)+	7 ldub [%i1+%g0],%g1	7 c=cMB[3]; r=rMB[3];	7 add %g3,%g4,%g4
ABS((int)c01-(int)r01)+	8 subcc %g2,%g1,%g4	8 SAD+=ABS((int)c-(int)r));	8 ...
ABS((int)c02-(int)r02)+	9 bpos 0x138		
ABS((int)c03-(int)r03);	10 ldub [%i1+(0x3)],%i2		
	11 sub %g1,%g2,%g4		
	12 and %i5,(0xff),%g2		
	13 and %g3,(0xff),%g1		
	14 subcc %g2,%g1,%g3		
	15 bneg,a 0x14c		
	16 sub %g1,%g2,%g3		
	17 add %g4,%g3,%g4		
	18 ...		

Fig. 5.16 Illustrating the impact of calculation reordering for the SAD application

afterwards for the calculation of the absolute difference results which are finally summed up. As presented in the assembly code, the compilation result follows this sequence as the memory loads are executed first and the calculations of the absolute values and the addition are done afterwards. For "*SAD* V3," the values are extracted pairwise and the absolute value and sum calculation is done immediately afterwards. Summarizing, the idea is to reduce the time the "*r*" and "*c*" values reside in registers and adding the result of the "absolute difference"-calculation earlier to the final result (i.e., the sum of absolute differences).

5.1.6   Selection of Transformations

Different transformations have different impact on different applications due to their diverse data and control flow. It is not necessary that every transformation will bring the best reliability benefit to for every application. Amongst these various transformed versions, the set of versions that are on the Pareto-optimal frontier should be selected. This way only the effective/appropriate function versions are selected per application. Tables 5.2, 5.3, and 5.4 present a general overview of the transformed reliable code versions that are selected for a given specific fault rate. Table 5.5 shows the detailed properties of different versions of the *SATD* application.

5.1.7   Impact on Critical ALU Instructions

Reliability improvement with reduced performance overhead can be achieved by employing reliability-driven software transformations. Conservatively applying redundancy to different parts of the code [27, 28, 77, 80, 92] may increase the spatial and temporal vulnerabilities, and thereby worsen the reliability. Furthermore, the total sum of the *critical instruction* executions becomes higher with the introduction of redundant *critical instructions*, which may ultimately lead to excessive rollbacks and increased performance/memory overhead due to an increased

Table 5.2 Best selected version for the case of 1 f/MCycles

Application	Performance of the baseline version [cycles]	Best selected version number	Data type	Unrolling	OM+CEE	Online table	Performance [cycles]
ADPCM	49481	V3	✓	✓	✗	✓	49222
CRC	2794826	V1	✓	✓	✗	✓	1740773
DCT	2385	V3	✗	✓	✓	✗	217
HT	2329	V3	✗	✓	✓	✗	164
IPRED	485	V5	✓	✓	✗	✗	224
MC-FIR	311	V3	✓	✓	✓	✗	119
SAD	2124	V4	✗	✓	✓	✗	2173
SATD	2306	V6	✓	✓	✓	✗	231
SHA	2765231	V2	✓	✓	✓	✗	1728195
SUSAN	2755035	V1	✗	✓	✓	✗	2405534

Table 5.3 Best selected version for the case of 5 f/MCycles

Application	Performance of the baseline version [cycles]	Best selected version number	Data type	Unrolling	OM+CEE	Online table	Performance [cycles]
ADPCM	49481	V3	✓	✓	✗	✓	49222
CRC	2794826	V1	✓	✓	✗	✓	1740773
DCT	2385	V3	✗	✓	✓	✗	217
HT	2329	V3	✗	✓	✓	✗	164
IPRED	485	V5	✓	✓	✗	✗	224
MC-FIR	311	V3	✓	✓	✓	✗	119
SAD	2124	V2	✓	✓	✓	✗	2373
SATD	2306	V5	✓	✓	✓	✗	229
SHA	2765231	V5	✗	✓	✓	✗	1698891
SUSAN	2755035	V1	✗	✓	✓	✗	2405534

Table 5.4 Best selected version for the case of 10 f/MCycles

Application	Performance of the baseline version [cycles]	Best selected version number	Data type	Unrolling	OM+CEE	Online table	Performance [cycles]
ADPCM	49481	V4	✓	✓	✗	✓	49481
CRC	2794826	V1	✓	✓	✗	✓	1740773
DCT	2385	V2	✗	✓	✓	✗	396
HT	2329	V3	✗	✓	✓	✗	164
IPRED	485	V5	✓	✓	✗	✗	224
MC-FIR	311	V4	✓	✓	✓	✗	124
SAD	2124	V4	✗	✓	✓	✗	2173
SATD	2306	V5	✗	✓	✓	✗	229
SHA	2765231	V5	✗	✓	✓	✗	1698891
SUSAN	2755035	V1	✗	✓	✓	✗	2405534

Table 5.5 Detailed example for SATD

	Cycles	#Executed instructions	Binary size (#instructions)	Data type	Unrolling	OM+CEE (%)	Online table
B	2306	1738	121	8 bit loads (100 %)	L1: 0	L2: 0	L3: 0	0	–
V1	704	528	168	8 bit loads (100 %)	L1: 4	L2: 2	L3: 2	100	–
V2	706	554	194	32 bit loads (100 %)	L1: 4	L2: 2	L3: 2	100	–
V3	425	313	136	8 bit loads (100 %)	L1: 4	L2: 4	L3: 4	100	–
V4	427	339	162	32 bit loads (100 %)	L1: 4	L2: 4	L3: 4	100	–
V5	229	197	197	8 bit loads (100 %)	L1: 4	L2: 16	L3: 16	100	–
V6	231	223	223	32 bit loads (100 %)	L1: 4	L2: 16	L3: 16	100	–

Fig. 5.17 Reduction of critical ALU instructions w.r.t. to baseline implementation for "SATD"

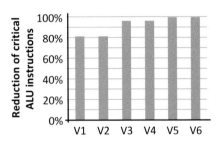

probability of Application Failures [28]. In order to alleviate this overhead, reliability-driven transformations can be employed that transform the instruction profile with reduced number of *critical instructions*, and thus reducing the error probability. Moreover, reducing the control flow and memory instructions also lowers the number of executions for critical ALU instructions used for address generation. These critical ALU instructions may cause Silent Data Corruptions. Figure 5.17 presents a detailed analysis for the *SATD* application. It illustrates that, compared to the baseline implementation, the proposed transformations reduce >80 % of the critical ALU instructions (i.e., address generation, condition instructions), which denote the predecessor instructions of critical memory and control flow instructions. It is important to note that the critical ALU instructions in the baseline implementation represent 58 % of the total arithmetic instructions in the *SATD* application, i.e., 732 critical ALU instructions out of 1248 total ALU instructions.

5.1.8 Impact on Performance Overhead When Employed Together with Error Detection and Recovery Techniques

The current state-of-the-art error detection and recovery techniques can be applied after applying reliability-driven transformations. This leads to a reduced performance overhead of applying redundancy. Figure 5.18 presents the performance analysis of instruction redundancy-based protection techniques *with* and *without* the proposed reliability-driven transformations. The analysis is normalized to the performance-optimized code. It is shown that employing these transformations in combination with the state-of-the-art instruction-redundancy-based techniques (like EDDI [27], SWIFT [28], and CRAFT [28]) offers up to 39.5 % (average 21.8 %) lower performance penalty compared to when not applying these transformations. One of the main reasons for the reduced performance penalty is the reduction in *branches* and *store* instructions that are used as synchronization points for inserting the check instructions in SWIFT and CRAFT techniques. In application programs with extensive *load/store* instructions (like *DCT, IPRED, SATD,* and *MC-FIR*), instruction redundancy-based techniques with the proposed transformations offer a considerable reduction in the performance overhead when compared to instruction redundancy without these transformations.

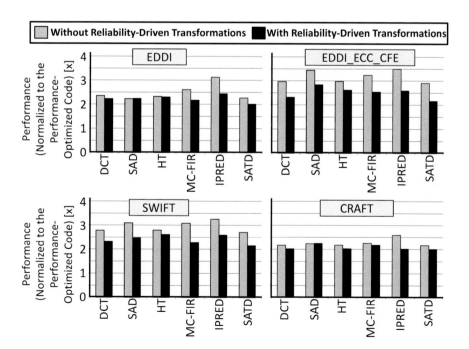

Fig. 5.18 Comparing instruction-redundancy techniques *with* and *without* the proposed reliability-driven software transformations

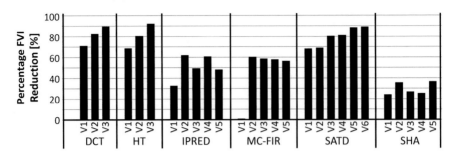

Fig. 5.19 FVI reductions of the reliability-optimized functions after applying the reliability-driven software transformations

5.1.9   Impact on FVI Reductions

The effects of transformations are also visible as FVI reduction in Fig. 5.19. The FVI takes both the Application Failures and Incorrect Outputs into consideration for the evaluation. The experiments show that the proposed transformations are effective to reduce the number of Application Failures, thus improving software reliability.

5.1.10   Summary of Reliability-Driven Transformations

In Sect. 5.1, four different reliability transformations are proposed, namely: (1) Reliability-Driven Data Type Optimization; (2) Reliability-Driven Loop Unrolling; (3) Reliability-Driven Common Expression Elimination and Operation Merging; and (4) Reliability-Driven Online Table Value Computation. These transformations reduce the number of *critical instruction* executions and spatial/temporal vulnerabilities. In this way, these transformations reduce the probability of errors. The detailed experimental evaluation demonstrates that these transformations lead to 60 % reduced Application Failures while reducing the FVI by 57 %. Furthermore, when employed in conjunction with different error detection and recovery techniques, the proposed transformations curtail the performance overhead of these techniques by 39 %. After employing reliability-driven transformations, the instruction vulnerabilities can be further reduced by optimizing the instruction execution sequence that affects the temporal vulnerabilities. Towards this end, a reliability-driven instruction scheduler is proposed in this work as discussed below in Sect. 5.2.

5.2   Reliability-Driven Instruction Scheduling

An instruction schedule determines the instruction execution sequence which can significantly impact the temporal vulnerability, i.e., vulnerable time that an instruction resides in different processor components, i.e., variable values in the register file or pipeline stage residency. An instruction scheduler optimizing only from the performance perspective may result in degraded reliability due to (1) increased program's susceptibility to Application Failures because a *critical instruction* may have been scheduled after a pipeline stalling instruction in order to increase the performance; and/or (2) increased spatial vulnerability because it used more registers or instruction execution units in order to achieve a higher performance. On the other hand, an instruction scheduler optimizing only for reliability may cause performance penalties due to *data hazards* that occur because of the data dependencies between different instructions, for instance, read-after-write, write-after-write, and write-after-read. The reliability-driven instruction scheduling becomes even more challenging in case of *structural hazards*[2] because the residency of instructions in the pipeline is longer due to frequent pipeline stalls and therefore increasing the probability of an Application Failure.

With the help of the following example scenarios shown in Fig. 5.20, we investigate these issues and motivate the need of considering both reliability and performance in instruction scheduling. The assumptions are: execution cycles (execute stage residency) of load/store=2 (two load-store units are available), multiply=3 (considering a dedicated multiply hardware unit), NOP=1.

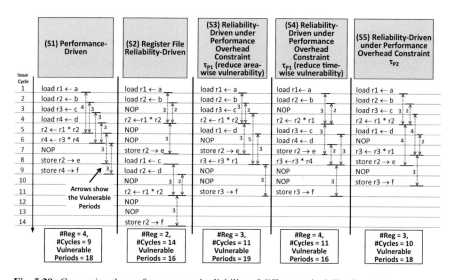

Fig. 5.20 Comparing the performance and reliability of different scheduling heuristics

---

[2] Occur due to resource conflicts in case the hardware cannot support simultaneous overlapped execution of instructions. For instance, only one register-file write port is available and the pipeline wants to perform two writes in the same cycle.

S1. **Performance-driven instruction scheduling (GCC list scheduler** [131, 132])[3]: This scheduler provides the best performance by prioritizing the instructions on the critical path of a software program. However, it does not take into account the spatial vulnerability (dependent upon the number of live registers) and the temporal vulnerability (i.e., their vulnerable periods) in its scheduling decision function. Consequently, it results in increased spatial and temporal vulnerabilities. The S1 schedule uses four registers and results in a total vulnerable period of 18. Furthermore, considering the *critical instruction* nature, scheduling four consecutive *loads* will result in an increased pipeline residency of critical instructions, e.g., due to a cache miss, and therefore a higher susceptibility towards Application Failures.

S2. **Register file reliability-driven instruction scheduling** [133]**:** The goal of this scheduler is to decrease the vulnerable periods between different register usages. Assuming the register allocation has been determined before instruction scheduling, it will also incur a reduced spatial vulnerability (only two registers are used). However, in order to avoid data hazards, NOPs are inserted that result in a performance overhead of five cycles (i.e., 55 %) compared to S1 (14 cycles of S2 vs. 9 cycles of S1). As a side effect, this schedule also avoids entering the *critical instructions* in the pipeline during the execution of multi-cycle instruction (*multiply*) or blocking instruction (*load*), thus it reduces the pipeline residency. In case a pipeline stalling instruction (e.g., a multi-cycle *multiply* instruction or a blocking *load* instruction) precedes a *critical instruction* (*load, store, address generation instructions*), the *critical instruction* will spend more time in the pipeline, i.e., increased pipeline residency. This results in an increased susceptibility towards Application Failures. Therefore, it might be beneficial to avoid scheduling a *critical instruction* right after an instruction that may potentially stall the pipeline.

S3. **Reliability-driven instruction scheduling under tolerable performance overhead** τ_{P1} **reducing only spatial vulnerability:** In contrast to S2, this scheduler minimizes the spatial vulnerability of instructions w.r.t. all processor components (and not only register file), while keeping the performance overhead under the tolerable limit ($\tau_{P1} = 20$ %). However, this schedule results in a significantly higher temporal vulnerability, i.e., a vulnerable period of 19.

S4. **Reliability-driven instruction scheduling under tolerable performance overhead** τ_{P1} **reducing only temporal vulnerability:** In contrast to S3, the goal of this scheduler is to optimize for temporal vulnerabilities within the tolerable performance overhead of $\tau_{P1} = 20$ %. It reduces the temporal vulnerability of variable "c" (stored in register $r3$) by shifting the *load* instruction right after the first *multiply* instruction. Finally, the S4 schedule results in a total vulnerable period of 16 with an execution time of 11 cycles (similar performance as of S3). The reduction in temporal vulnerability comes at the cost of

---

[3] Here, the Haifa List Scheduler of GCC is used, where an instruction gets into the ready queue once the dependencies are resolved and all the resources to execute this instruction are available. In GCC, the instruction scheduling is done in a two-pass fashion; the second pass is applied after the register allocation stage.

an increased spatial vulnerability (four registers are used in S4 instead of three registers used in S3). However, as a side effect, the program's susceptibility towards Application Failures is increased due to the scheduling of the *critical instructions* like *load* and *store* right after the multi-cycle *multiply* instruction.

S5. **Reliability-driven instruction scheduling under tolerable performance overhead τ_{P2}:** Unlike S3 and S4, a joint optimization for both the spatial and temporal vulnerabilities is performed under a given tolerable performance overhead of $\tau_{P2} = 10$ % (i.e., even a tighter performance overhead constraint compared to S3 and S4). The spatial and temporal vulnerabilities are reduced compared to S4 and S3 schedules, respectively, while achieving a performance close to the performance-driven S1 schedule. In this schedule, it can be seen that the *load* and *store* instructions are not be scheduled after the *multiply* instruction unlike S4. This results in a reduced susceptibility towards Application Failures. Amongst the five schedules, S5 provides a good compromise between performance overhead and reliability improvement.

Summarizing the Above Scenarios: Since an instruction scheduling algorithm determines the execution sequence of instructions, it also affects the spatial and temporal vulnerabilities of instructions in different processors components. A performance-driven instruction scheduler may result in reduced temporal vulnerability for instructions that lie on the critical path; however, it may lead to a significantly higher spatial vulnerability by using more resources or even longer vulnerable periods of the operands (inside the register file) of the *critical instruction* if they do not lie on the critical path. Therefore, it may lead to an overall higher vulnerability. On the other hand, a reliability-driven instruction scheduler would not only *balance the spatial and temporal vulnerabilities*, but would also *account for critical instructions* such that the software reliability is improved, though it may incur a performance loss. Therefore, under a user-provided tolerable performance overhead, a reliability-driven instruction scheduler needs to optimize the software reliability which is a complex function of spatial and temporal vulnerabilities of various instructions w.r.t. different processor components/resources utilized during their execution. Finally, a reliability-driven instruction scheduler needs to incorporate the knowledge of *critical* and *noncritical instructions* in order to reduce the program's susceptibility towards Application Failures.

5.2.1   Soft Error-Driven Instruction Scheduling

Towards this end, a *soft-error driven instruction scheduler* is proposed in this manuscript that schedules the instructions with the objective to enhance a software program's reliability against soft errors under a user-provided tolerable performance overhead. As mention in Sect. 3.2, the instructions are classified as *critical* and *noncritical* depending upon the severity of potential program errors due to the hardware level faults. As a comprehensive reliability cost function, the proposed soft-error-driven instruction scheduler incorporates the idea of *instruction reliability weight*, which is a joint cost function of the statically estimated instruction vulnerability (spatial and temporal vulnerabilities of various processor components used during the execution of a certain instruction) along with instruction's criticality, probabilities of different error types, and number of dependent instructions. The scheduler employs a *lookahead*-based heuristic for evaluating the reliability weights of various scheduling candidate instructions while taking into account the reliability weights of the successor instructions. The *lookahead* property reduces the risk of scheduling a *critical instruction* after a pipeline stalling instruction.

The proposed scheduler operates at the basic block level and the input is an instruction dependency graph with the estimated IVI of each instruction, processor model, and the predicted probability of execution of each basic block. It outputs a reliability-enhanced assembly code with a soft-error-driven instruction schedule. The scheduler employs a *lookahead*-based instruction scheduling heuristic that maximizes the reliability weight of a basic block. In the following, the input, output, and optimization goal are presented formally, followed by the novel scheduling heuristic.

5.2.2   Formal Problem Modeling

Input: Given a Basic Block (BB) as a directed acyclic graph $G = (V, E)$, where V is a set of nodes denoting the instructions, such that $V = \{n_1, \ldots, n_N\}$, where N is the number of instructions in the BB. Each node is given as a tuple $n_i = \{S_i, P_i, IVI_i, T_i, eT_i, d_i, o_i\}$. T_i, eT_i, d_i, $o_i = \{o_1, \ldots, o_m\}$ are instruction execution time, earliest time an instruction can be scheduled, and destination/source operands, respectively. For a node i, $S_i = \{s_1, \ldots, s_x\}$ and $P_i = \{p_1, \ldots, p_y\}$ represent the sets with successor and predecessors nodes, respectively. E is a set of edges denoting the instruction dependencies, such that $E = \{e_{ni \to nj} \mid n_i, n_j \in V\}$, where $e_{ni \to nj}$ is the latency of moving from instruction n_i to n_j. The execution frequencies and probabilities of basic blocks are predicted using the GCC framework [121, 129].

Output: A reliability-optimized instruction schedule G_S.

Constraint: A user-provided tolerable performance overhead P_τ, i.e., tolerable performance loss compared to the performance-driven instruction schedule. In this work, we use the instruction scheduler of the GCC framework as a basis. It schedules the instruction with the maximum delay (δ_{MAX}) first.

Optimization Goal: An *Instruction Reliability Weight* (ψ) is employed as the objective function, such that the *instruction which has the highest reliability weight* (ψ_{MAX}) *is scheduled first* in case the performance loss introduced by scheduling that instruction is below P_τ. In addition to the *estimated Instruction Vulnerability Index* (*IVI*), ψ incorporates the *criticality* of an instruction that corresponds to the Application Failures and the number of dependent instruction in the same and subsequent dependent basic blocks as discussed below.

Program Error Types: Considering the type of an instruction, i.e., *critical instruction* (CI) or *noncritical instruction* (nCI), probabilities of an Application Failure, and an Incorrect Output (P_{Failures}, $P_{\text{IncorrectOP-CI}}$, $P_{\text{IncorrectOP-nCI}}$) are employed in the ψ computation as shown in Eqs. 5.2 and 5.3. These probabilities are obtained using fault injection experiments at a certain fault rate "f".

$$\psi_{nCli} = \left(\prod_{\forall c \in \text{Proc}} P_{IncorrectOP-nCI}(c) \right) \times IVI_i \qquad (5.2)$$

$$\psi_{Cli} = \left(1 + \prod_{\forall c \in \text{Proc}} P_{Failures}(c) + \prod_{\forall c \in \text{Proc}} P_{IncorrectOP-CI}(c) \right) \times IVI_i \qquad (5.3)$$

Dependent Instructions: The reliability weight of an instruction ψ also depends upon the number of dependent/successor instructions (I_D) in the same basic block (BB, with N instructions) because an instruction with several successors exhibits a higher potential for fault propagation (see Eq. 5.4). It illustrates that an instruction might become (reliability-wise) important if it has dependent *critical instructions*. Some dependent instructions may also exist in other basic blocks. It may result in fault propagation to those dependent basic blocks (DB). In this case, the goal is to keep such an instruction near the end of the basic block. This incurs a negative cost (Eq. 5.5). This observation motivated a *lookahead*-based instruction scheduling heuristic.

$$\psi_i' = \sum_{\forall j \in \{CI, nCI\}} \psi_{ji} \times I_{Dj} / N \qquad (5.4)$$

$$\psi_i = \psi_i' - \sum_{\forall j \in \{CI, nCI\}} \psi_{ji} \times \left(\sum_{\forall k \in DB} I_{Djk} / N_k \right) \qquad (5.5)$$

The reliability weights of the complete basic block and software function are quantified as *basic block reliability weight* (ψB) and *function reliability weight* (ψF), respectively.

$$\psi B = \sum_{i=1...N} \psi_i / N \qquad (5.6)$$

$$\psi F = \sum_{\forall j \in \{BB\}} \left(\sum_{k=0}^{nExec_j} \psi B_{jk} / T_F \right) \qquad (5.7)$$

$nExec_j$ is the number of predicted executions of BB_j and T_F is the execution time of the function "F".

5.2.3 Lookahead Instruction Scheduling Heuristic

The lookahead property is incorporated in the soft-error-driven instruction scheduling heuristic. Considering this property, the heuristic evaluates the reliability weight of an instruction in conjunction with the reliability weights of its successor instructions. In addition to this, it keeps track of the performance loss compared to a performance-driven instruction scheduler.

An Example: Figure 5.21 presents an abstract example of the lookahead scheduling concept. Let an instruction node "*a*" be scheduled at level "*j*−1". The typical choices for the next instructions at level "*j*" would be instruction nodes "*b*", "*c*", and "*d*". However, it might be reliability-wise beneficial, if another level "*j*+1" is explored. This is because, the dependent instructions of, for instance, "*b*" are *critical instructions* and "*c*" and "*d*" are *noncritical instructions*. Considering this, the soft-error-driven schedule leads to a schedule of "+, +, +, *, *, store" (see example scenario in Fig. 5.21). However, scheduling the left-side *add* instruction after the right-side *add* and scheduling *store* after the *multiply* instruction might raise two reliability concerns: (1) the vulnerable period of the *store* input operand is increased, (2) the vulnerable period of the address value of the *store* instruction is significantly longer. Therefore, when jointly taking into account the reliability at two scheduling levels "*j*" and "*j*+1", the following schedule is obtained "+, +, store +, *, *", where the left-side *add* and *store* instructions are scheduled before the right-side *add*. Compared to the first schedule, the second schedule exhibits a higher reliability weight (ψB). In the above example scenario, only two levels are explored by the lookahead heuristic. However, the proposed lookahead concept is generic and further levels can also be explored, but that would exponentially increase the scheduling complexity. Therefore, exploring two levels in the lookahead was adopted as a compromise between complexity and reliability improvement.

Considering the above example and lookahead for two levels, the *optimization goal* is:

$$Maximize\left[\left(\psi b + \max_{i=1,...,l} \psi b_i\right), \left(\psi c + \max_{j=1,...,m} \psi c_j\right), \left(\psi d + \max_{k=1,...,n} \psi d_k\right)\right] \quad (5.8)$$

Due to the lookahead into the subsequent scheduling levels, the instruction scheduler avoids selecting locally best solutions at a certain scheduling level. It is due to the fact that the reliability weights of the dependent instructions on the subsequent scheduling levels are also incorporated in the optimization goal.

Lookahead Heuristic: Figure 5.22 presents the flow of the lookahead heuristic for instruction scheduling (see detailed algorithm in Appendix C). First, the sets of

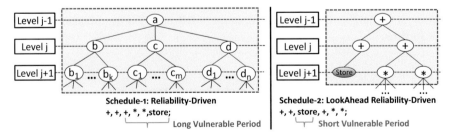

Fig. 5.21 An example of the soft-error-driven instruction scheduler with lookahead heuristic

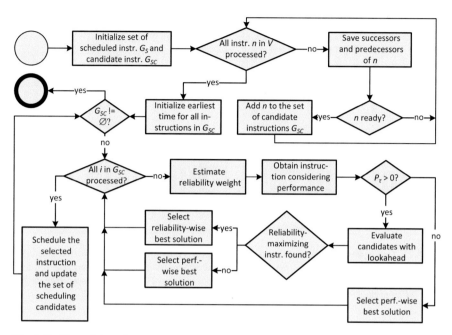

Fig. 5.22 Flow of the soft-error-driven instruction scheduler

scheduled and scheduling-candidate instructions are initialized. The successor and predecessor instructions for each instruction are obtained and the ready list is initialized. Afterwards, initialization of earliest time and current time is performed. The process iterates until the ready list is empty and all scheduling candidates are processed. As shown in Fig. 5.20, the vulnerable periods of instructions depend upon the scheduled instruction. Therefore, in every iteration, the reliability weights of all instructions are re-estimated. If within the tolerable performance overhead constraint, the instructions in the candidate set are evaluated for reliability using a *lookahead* approach that selects an instruction with the highest value of the sum of the reliability weights of the instruction and its successor instruction. Otherwise, the performance-wise best solution is adopted. Afterwards, the sets of scheduled instructions and scheduling candidates are updated accordingly considering the resolved dependencies.

5.2.4 Results for the Reliability-Driven Instruction Scheduling

Comparison of Error Distribution Results with State-of-the-Art Scheduling Techniques: Figure 5.23 presents the comparison of the error distribution of the proposed reliability-driven instruction scheduler (under three different tolerable performance overhead constraints, for three different fault rates) with the performance-driven scheduler [131] and state-of-the-art reliability-driven

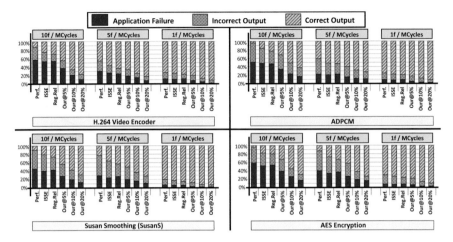

Fig. 5.23 Comparing the error distribution of the proposed reliability-driven instruction scheduler (at three different tolerable performance overhead constraints) with the performance-driven and state-of-the-art ISSE [81] and register file reliability improving [133] instruction schedulers for three different faults rates

schedulers ISSE [81] and register file reliability improving scheduler [133]. When comparing with the state-of-the-art [81] and [133], the proposed reliability-driven scheduler offers reduction in Application Failures of up to 15 % at 5 % performance loss and up to 45 % at 20 % performance loss. The reason is that it incorporates the knowledge of *critical* and *noncritical instructions* in the reliability weight. Specifically, the reduction in Application Failures is because of a lower number of corrupted address operands of the *critical instructions* (*load/store* instructions) due to reduced temporal vulnerability of these *critical instructions* that are not on the critical path. Compared to the scheduler of [133], the proposed scheduler achieves a reduced number of Application Failures as the pipeline residency of *critical instructions* is reduced. The primary reason of performance loss is because, the spatial vulnerability is reduced at the cost of higher temporal vulnerability, and this results in the overall vulnerability reduction.

It is evident that, when moving from 5 to 10 % and to 20 % performance loss cases, the number of Application Failures are reduced. In case of *ADPCM* and *AES* applications, when moving from 10 to 20 % performance loss, there is a small reliability improvement which is attributed to the fact that both the spatial and temporal vulnerabilities are balanced. Furthermore, as the tolerable performance overhead increases, the percentage of Application Failures decreases due to the lookahead nature, that prefers scheduling a *critical instruction* of the upcoming scheduling level first in order to reduce its vulnerable periods. However, this may result in a slight increase in the percentage of Incorrect Outputs, which may be less severe compared to the Application Failures. Overall, the proposed scheduler offers an increase in the correct output cases by average 22 % (averaged over various tolerable performance overheads), when compared to state-of-the-art instruction schedulers [81, 133]. For fairness, all schedulers were evaluated using the same reliability model.

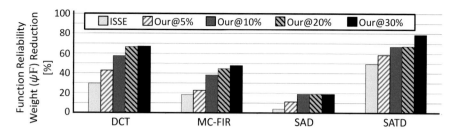

Fig. 5.24 Comparing the function reliability weight reductions to ISSE [81]

Analyzing the Impact on Function Reliability Weight: Figure 5.24 shows the reduction in the function reliability weight (ψF) of the proposed reliability-driven instruction scheduler (under different tolerable performance overheads) and ISSE [81] compared to a performance-driven instruction scheduler [131] for different kernels of the *H.264* video encoder application. The proposed scheduler reduces the ψF value by up to 79 % (on average 53 %) at the cost of a 30 % performance loss. For a 5 % tolerable performance loss, the proposed scheduler reduces the ψF value by up to 58 % (on average 33 %). Note, when moving from 20 to 30 % tolerable performance loss, in some applications (*DCT, MC-FIR*) there is limited reliability improvement, because the potential for reliability improvement at the software level for these applications has already been exploited. In case of *SAD*, because of the excessive load instruction executions, the reliability improvement potential is limited. Therefore, it is fully exploited for a 5 % tolerable performance, and there is insignificant reliability improvement by the instruction scheduler after the 10 % case. Still in these cases, the reliability improvement of the proposed scheduler is better when compared to that of ISSE [81]. For large-sized programs (like *SATD*), the proposed scheduler offers consistent reliability improvements.

5.2.5   Summary of Reliability-Driven Instruction Scheduling

A reliability-driven instruction scheduler is presented that employs a lookahead-based heuristic to schedule instructions under tolerable performance overhead constraints. It evaluates the reliability of dependent instructions considering the impact of spatial and temporal vulnerabilities w.r.t. various processor components. The proposed reliability-driven instruction scheduler provides on average a 22 % reduction of Application Failures compared to state-of-the-art. Furthermore, it reduces the function vulnerability by 33–53 % on average compared to the state-of-the-art ISSE technique [81]. The above-discussed reliability-driven transformations and instruction scheduling reduce the error probabilities, especially related to control flow and memory related instructions, thus leading to a reduced number of Application Failures. However, for data computing instructions, there is still a need to apply instruction redundancy to obtain a high degree of software reliability. Therefore, this work also proposes a novel *selective instruction redundancy technique* that leverages a detailed program analysis to identify error propagation paths in the data and control flow along with vulnerabilities, error masking, and error propagation probabilities. It then applies instruction-level redundancy on selected reliability-wise important instructions under a user-provided tolerable performance overhead constraint. Such a technique is complementary and orthogonal to the reliability-driven transformations and instruction scheduling, and can be applied in conjunction with them to achieve a high degree of reliability.

5.3   Reliability-Driven Selective Instruction Protection

The instruction-level Error Masking Index (IMI) and Error Propagation Index (EPI) capture the error masking and propagation properties of the software program. These models are used to *enable constrained software reliability optimization on unreliable hardware* considering software-level error detection and recovery feature. These models are used for prioritizing the instructions in basic blocks with respect to reliability and facilitate the tradeoff between performance loss and reliability improvement. Towards this end, a selective instruction protection technique is proposed that selectively protects instructions or group of consecutive instructions in different execution paths in a given function. For this, it leverages the *EPI* and *IVI* to compute the instruction reliability profit.

5.3.1   Reliability Profit Function for Choosing Instructions for Protection

The proposed instruction protection heuristic employs a *reliability profit function* (RPF, Eq. 5.9) which is defined as the accumulated *reliability efficiency* of a group of instructions g, such that the reliability efficiency is given as the product of *EPI* and *IVI* divided by the protection overhead ω. The overhead of the instruction group g is computed using Eq. 5.9; *csi* is the set of consecutive instructions and *ci* is the checking instruction inserted only at the end of g or at the point of multiple outputs.

$$RPF = \left(EPI(g) \times IVI(g)\right)/\omega_g; \quad s.t., \quad \omega_g = \sum_{\forall csi \in g} t_{csi} + \sum_{\forall ci \in g} t_{ci} \qquad (5.9)$$

A Comprehensive Example Comparing Different Potential Solutions: Figure 5.25 shows an excerpt from an example instruction graph with 13 instructions and the effect of different parameters in the optimization goal on the efficiency of selective instruction protection under a tolerable overhead budget of 20 cycles. The table in Fig. 5.25 provides *IVI*, *EPI*, and ω of each instruction *I*.

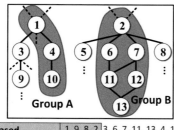

I	IVI(I)	EPI(I)	ω(I)
1	1.0	4.0	4
2	0.90	3.5	3
3	0.85	3.5	8
4	0.65	2.5	3
5	0.40	2.0	5
6	0.85	2.5	2
7	0.85	3.0	3
8	0.95	1.5	8
9	0.95	3.3	5
10	0.65	1.7	2
11	0.85	2.5	3
12	0.60	2.0	2
13	0.80	1.8	3

(a) IVI-Based | 1, 9, 8, 2, | 3, 6, 7, 11, 13, 4, 10, 12, 5
(b) (IVIxEPI)-Based | 1, 2, 9, 3, | 7, 6, 11, 10, 13, 8, 4, 12, 5
(c) (IVIxEPI)/ω-Based | 6, 2, 1, 7, 11, 9, | 12, 10, 4, 13, 3, 8, 5
(d) Group-Based | 2, 6, 7, 11, 12, 13, 1, 4, 10, | 3, 9, 8, 5

Values in the filled-block area denote the **protected** instructions
Overhead Budget: 20 Cycles

Fig. 5.25 An example showing the effect of different parameters on the reliability efficiency of the selective instruction protection

IVI-Based Selection: Figure 5.25a illustrates that when only considering the instruction vulnerabilities, four instructions are chosen for protection. However, there can be cases, where an instruction's vulnerability to error is quite high, but the probability that this error will be masked until the visible output is also high.

(IVI × EPI)-Based Selection: Figure 5.25b shows that when jointly considering the *IVI* and *EPI* metrics to evaluate the instructions for their reliability-wise importance, instruction 3 is chosen instead of instruction 8. However, still a total of four instructions are protected (like previous case) due to the high protection overhead of instruction 3 and 9. In such scenarios, it might be beneficial to choose several instructions with a slightly lower *IVI×EPI* profit and low protection overhead, rather than protecting only few instructions with high protection overhead. Note, depending upon the instruction types and protection mechanism, the protection overhead may vary significantly for different instructions.

Instruction Reliability Efficiency-Based Selection: Figure 5.25c illustrates that six instructions are protected, while the total reliability efficiency is 0.854 (computed using Eq. 5.9 and values in the table), which is 46 and 29 % better compared to the case (a) having a reliability efficiency of 0.585 and case (b) having a reliability efficiency of 0.663, respectively. Note, the protection overhead depends upon the protection mechanism. In case, the underlying protection mechanism is simple software level error detection (like SWIFT [28]) or recovery (like SWIFT-R [71]), the checking or voting instructions are incorporated at the store instructions or leaf nodes of the groups. Therefore, protection overhead may be curtailed by computing the group reliability efficiency, i.e., cumulative reliability efficiency for a group of consecutive instructions.

Group Reliability Efficiency-Based Selection: Figure 5.25d illustrates the marked regions in the graph as groups of protected instructions. Note, using group reliability efficiency, nine instructions are protected, while the overall reliability efficiency is 0.966, which is 13 % better compared to that of the case (c).

5.3.2 Flow of the Selective Instruction Protection Heuristic

Figure 5.26 shows the flow of the proposed selective instruction protection heuristic (see detailed algorithm in Appendix C) that protects a set of reliability-wise critical instructions in a given instruction graph G, using a user-defined reliability method R, under a user-provided tolerable performance overhead budget P_τ. It aims at maximizing the total reliability profit function of a group of instructions to avoid excessive checking/voting instructions. First, the reliability profit function for each instruction is individually computed and inserted into a list that is then sorted in the descending order. Since the generation of all groups of all instructions leads to a significant complexity, the heuristic starts with protecting individual instructions and incrementally builds instruction groups for protection considering the predecessor and successors of the protected instructions. Afterwards, it re-estimates their combined protection overhead and inserts the check/voting instructions at the appropriate locations, e.g., group boundaries. The reliability profit function is computed for all the groups and the list of instruction is re-sorted such that the instructions of current best group with the highest reliability profit appear first in the list, which is later evaluated for protection.

Note, it is assumed that the control flow is protected using standard techniques like basic block signature [34]. This work employs SWIFT-R [71] as the basic protection mechanism. However, the proposed technique for constrained program reliability optimization and models for error propagation and masking are equally beneficial for selective applicability of other protection mechanisms and orthogonal to improvements in such program-level recovery mechanisms.

5.3.3 Results for Selective Instruction Protection

The proposed selective instruction protection technique for constrained program reliability optimization is compared for reliability efficiency with various state-of-the-art program-level protection techniques [71, 76, 134] (Fig. 5.27b–d) and the

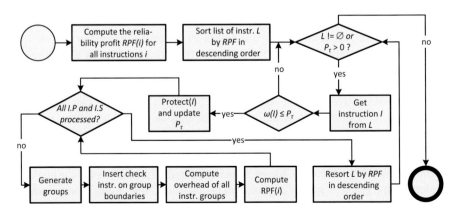

Fig. 5.26 Flow of the selective instruction protection technique

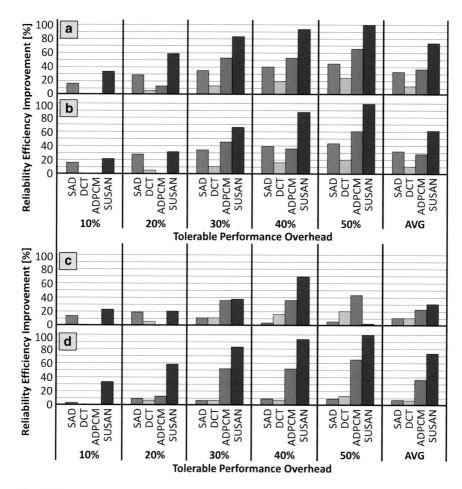

Fig. 5.27 Comparing the reliability efficiency improvement of the proposed selective instruction protection technique over (**a**) unprotected case; and three state-of-the-art techniques namely (**b**) SWIFT-R [71] under constraint; (**c**) instruction vulnerability factor (IVF)-based selective protection [76]; and (**d**) instruction dependency-based selective protection [134]

baseline unprotected case (Fig. 5.27a) for varying number of tolerable performance overhead cases for different applications. For fairness, all comparison partners are evaluated with same fault scenarios, same compiler options, thus same application binaries, control and data flow graph, same input data, and same basic protection mechanism, i.e., instruction level TMR and voting. The results only represent the effect due to difference in the protection cost function and selection technique.

Overall Comparison with all State-of-the-Art Techniques: First, the experimental observations that are common in all comparisons are discussed. For the *SusanC* application, the proposed selective protection technique obtains a high reliability improvement (average 30–60 %) over all comparison partners because the *SusanC* application exhibits instructions with high IMI and EPI indexes having varying distribution.

Due to the joint consideration of error propagation/masking probabilities and vulnerability in the cost function (Eq. 5.9), the proposed technique stays superior compared to all state-of-the-art protection techniques and achieves a reliability efficiency improvement of up to 60–99 % (average 30–60 %). For 50 % tolerable overhead, the reliability of the *SusanC* application reaches close to 100 %, because all the important non-masking instructions are protected within this tolerable overhead budget, while masking instructions are left un-protected as errors during these instructions do not affect the correct program output. This illustrates the benefit of the proposed selective protection technique since it accounts for error masking and propagation probabilities in the protection cost function. The improvements are also high in case of the *ADPCM* application for overhead cases of 30 % and higher (average 30–40 % reliability efficiency improvement). Below 20 % the efficiency is low, as important instructions require more overhead than the tolerable overhead budget due to their high execution frequency. The improvements in the *SAD* application are also noticeable, up to 45 % and average 10–30 % reliability efficiency improvement. However, the improvements for the *DCT* application are relatively low, due to limited masking probability of instructions and dependency on the earlier instructions of the algorithm, on average 8–10 % reliability efficiency improvement. For low overhead cases, for several applications like in *DCT* and *SAD*, the savings of the proposed technique are below 10 %. This is because of the fact that important instructions, that typically occur in loops, have many executions and their required protection overhead cannot be fulfilled under the cap of 5 or 10 % tolerable performance overhead. In the following, specific observations for different comparison cases are discussed.

Comparing with the Unprotected Case: Comparison with the unprotected case shows the best possible reliability efficiency improvement of the proposed selective protection technique. The proposed technique offers up to 25–99 % and average 30–70 % improvement in the reliability efficiency compared to the unprotected case for different applications.

Comparing with the SWIFT-R Technique [71]: SWIFT-R [71] technique is the most prominent program-level instruction protection technique that employs TMR with majority voting for protecting all instructions. Compared to the original SWIFT-R (which is unconstrained to performance overhead), the proposed technique achieves more than 3× better reliability efficiency, since the overhead of SWIFT-R is more than 5×–6×. For fairness of comparison, the SWIFT-R implementation is adapted towards constrained optimization, such that the overhead constraint is used to determine the number of instructions that can be protected. Afterwards, the instructions are selected for protection in a sequential manner, i.e., first execute, first protect. The proposed technique offers up to 20–97 % and on average 10–60 % improvement in the reliability efficiency compared to the constrained SWIFT-R variant.

Comparing to the Instruction Vulnerability Factor-Based Selective Protection Technique [76]: This technique computes the instruction vulnerability and protects the instruction first which has the highest vulnerability factor. However, this technique does not account for the instruction dependencies and error propagation properties. Therefore, it only works well for cases where vulnerability is dominant and the error propagation is very low with smooth distribution. In contrast, the proposed selective protection technique accounts for both vulnerability and error propagation properties. As a result, the proposed technique provides up to 20–70 % (average 10–30 %) reliability efficiency improvement compared to the Instruction Vulnerability Factor-based protection technique mentioned in [76].

Compared to the Instruction Dependency-Based Selective Protection Technique [134]: This technique prioritizes and protects the instructions which have more dependent instructions. However, it ignores the error masking probabilities and instruction vulnerabilities. As a result, in some cases like for the *SusanC* application, the technique proposed in [134] provides significantly less protection. However, in cases like for the *DCT* application where error propagation is crucial and dominant over vulnerability, this technique provides good reliability. In contrast, the proposed selective protection technique provides high reliability efficiency in all cases, as it jointly accounts for error propagation and masking probabilities, vulnerabilities, and overhead of different instructions individually or jointly in a group. The proposed protection technique thereby achieves up to 12–99 % (average 7.5–80 %) improved reliability efficiency compared to the technique of [134].

5.3.4 Summary of Selective Instruction Protection

In Sect. 5.3, a novel selective instruction protection technique for software program reliability optimization under tolerable performance overhead constraints is proposed. This technique exploits program-level error masking and propagation properties to perform reliability-driven prioritization of instructions and selective protection during compilation. This leverages the statistical models for estimating error masking and propagation probabilities. The proposed selective protection technique provides significant improvement in reliability efficiency (on average 30–60 %) compared to different state-of-the-art program-level protection techniques.

5.4 Multiple Function Version Generation and Selection

Different functions of the same application exhibit distinct instruction profile and control flow resulting in unique reliability and performance properties, as shown by the varying FVI and execution time values in Fig. 5.28a. Moreover, the same function when implemented using different algorithms exhibits distinct performance and reliability properties. An example in Fig. 5.28b illustrates that different sorting algorithms and even different implementations by different programmers have different FVI and execution time values. Similarly, enabling different transformation options may also result in significant impacts on the FVI and execution time of the same function. The goal is to select a representative set of function versions to cover a wide range of possible FVIs and average execution times for the resulting binary code of a function version.

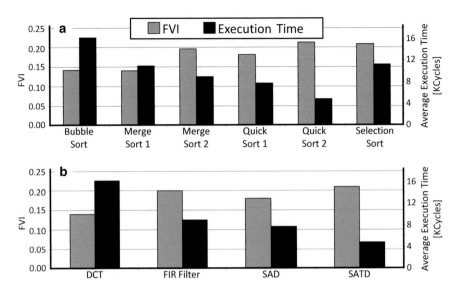

Fig. 5.28 Different algorithms have different FVI and average execution time: (**a**) comparing different sorting algorithms; (**b**) comparing different functions of the same application

In the following, the flow for generating up to K_i versions of binary codes is explained. Given are the function implementations, the code transformation methods, instruction scheduling method, and selective instruction redundancy technique under a user-specified tolerable performance overhead. Here, the tolerable performance overhead is defined as an upper bound of the increase in the average-case execution time, compared to the best performance version under the average-case execution time. As discussed in earlier sections, it is usually better to first adopt the transformation methods to explore potential reliability improvement. Following that, these transformation methods are adopted to obtain the corresponding *FVI* and average execution time. Amongst the obtained binary translations, the set of binary version implementations for *RTP Pareto-frontier* is taken, i.e., none of any two binary version implementations will dominate each other in both *FVI* and the average execution time. Then, amongst all the binary version implementations, the instruction rescheduling method is further considered to additionally exploit some local improvement. The set of binary version implementations for *RTP Pareto-frontier* is updated. These two reliability-driven transformation steps reduce the error probability. Afterwards, the selective instruction redundancy is applied for software-level error detection and recovery. Figure 5.29 illustrates different selected versions with reliability-performance tradeoffs. Finally, the obtained pareto-optimal functions versions are forwarded as an input to the reliability-driven system software for dependable execution of application software programs.

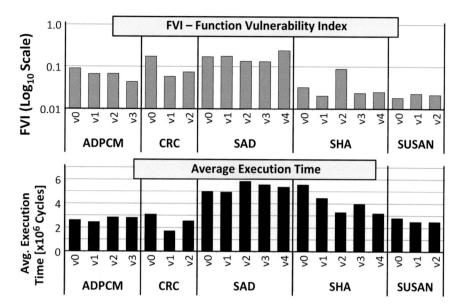

Fig. 5.29 Performance and vulnerability of different compiled versions of different applications

5.5   Chapter Summary

Overall the proposed reliability-driven transformations, instruction scheduling, and selective protection techniques comprehensively exploit the instruction-level vulnerabilities, error masking and error propagation properties along with the knowledge of critical and noncritical instructions and their interdependencies. These transformations reduce the number of critical instruction executions and spatial/temporal vulnerabilities, thereby lowering the error probability during the application execution. The reliability evaluations of these transformations demonstrate a reduction of 60 % in Application Failures and 57 % in the overall vulnerability compared to a performance-optimized code. However, these transformations do not provide error detection and recovery features. Therefore, additional techniques are required for instruction-level redundancy to achieve error detection and recovery in order to ensure further reliability improvements. Towards this end, it is also analyzed that conservatively applying redundancy to the code without the information of the varying reliability properties of different instructions may increase the total number of critical instruction executions ultimately leading to increased Application Failures, excessive rollbacks and increased performance/memory overhead. When these transformations are employed in conjunction with different error detection and recovery techniques, the proposed transformations curtail the performance overhead of these techniques by 39 %. Furthermore, the reliability-wise impact of these transformations is distinct for different applications due to the varying potential attributed to applications' diverse data and control flow. Therefore, amongst various transformed versions, the set of versions that are on the Pareto-frontier are selected. Afterwards, a reliability-driven lookahead-based instruction scheduler is employed that schedules instructions by evaluating the reliability of dependent instructions while reducing spatial and temporal vulnerabilities and considering the knowledge of the critical and noncritical instructions. The proposed reliability-driven instruction scheduler reduces Application Failures by 22 % on average compared to state-of-the-art instruction schedulers. Since the above-discussed transformations and instruction scheduler only reduce the error probabilities, to achieve further reliability improvements, additional instruction-level protection techniques are developed that deploy selective instruction redundancy for error detection and correction in constrained scenarios. The selective protection technique jointly accounts for instruction-level vulnerability, masking, and error propagation knowledge to prioritize different instructions and selects a group of most reliability-wise beneficial instructions for applying redundancy under a tolerable performance overhead constraint. The proposed technique offers significant improvement in reliability efficiency (on average 30–60 %) compared to different state-of-the-art program-level protection techniques. In short, the above-discussed software program-level reliability optimization techniques for dependable code generation supersede state-of-the-art software-level reliability techniques.

The software transformations and selective instruction redundancy techniques presented in this chapter aim at enhancing software program's reliability under tolerable performance overhead constraints. State-of-the-art techniques try to incur no performance overhead or provide full redundancy with a very high overhead, whereas the techniques presented in this manuscript leave it up to the user to specify a tolerable performance overhead that may be associated with a gain in reliability. Often, a small performance loss may enable a high reliability gain. The proposed techniques can exploit those scenarios, state-of-the-art techniques do not.

The techniques proposed in this chapter are leveraged in a reliability-driven compilation flow in order to generate multiple reliable versions for different application programs/functions that enable various tradeoff options between reliability and performance. These multiple reliable function versions are leveraged by other system layers, for instance, by offline and online system software as shown in Chap. 6, to optimize the reliability under constrained scenarios.

Chapter 6
Dependable Code Execution
Using Reliability-Driven System Software

The multiple compiled function versions generated in Chap. 5 are leveraged by the reliability-driven system software to exploit the vulnerability vs. performance trade-offs for optimization under constrained scenarios. This chapter presents several system software level reliability optimization techniques in order to enable dependable execution of applications on single core processors and multi-/manycore processors, while exploiting the concept of multiple reliable versions. In case of timing-conscious systems, besides functional correctness, it is also important to consider the timing correctness (i.e., whether a correct output is delivered within the deadline or not). To account for both functional reliability (i.e., correctness of program's output in the presence of faults) and timing reliability (i.e., deadline misses in the presence of faults, for instance, due to reliable application execution with performance loss), the system software layer employs the *Reliability-Timing Penalty* as the optimization function. First, a novel offline system software is presented in Sect. 6.1 that constructs a schedule table for different function versions such that these schedules offer minimum Reliability-Timing Penalty. Each schedule represents a particular dependable application composition under a timing scenario. At run-time, in order to provide dependable application execution for single core processors, a reliability-driven run-time system (Sect. 6.2) selects appropriate reliable versions depending upon the execution properties of different functions, deadline, and the achieved reliability levels.

In case of multi-/manycore systems, further reliability improvements can be obtained by leveraging the architectural support for redundant multithreading. However, ensuring high soft error resilience in manycore processors needs to account for other reliability threats (like process variations and aging), too. Since process variations and aging induce core-to-core frequency variations, employing redundant multithreading will require synchronization of the outputs of redundant threads executing on different cores. This may potentially lead to deadline misses, which are undesirable. State-of-the-art redundant multithreading techniques ignore the core-to-core frequency variations and consider chip-level guardbanding, which results in a significant loss in the performance potential of the overall chip.

© Springer International Publishing Switzerland 2016
S. Rehman et al., *Reliable Software for Unreliable Hardware*,
DOI 10.1007/978-3-319-25772-3_6

Furthermore, in resource-competing or resource-constrained scenarios, it may not be possible to execute all of the concurrently executing applications in the redundant multithreading mode. Therefore, it may be important to prioritize applications depending upon their functional and timing reliability. Towards this end, this chapter introduces a novel *Dependability Tuning* system for manycore processors (Sect. 6.3) that manages reliable execution of multiple concurrently executing applications considering support for redundant multithreading, core-to-core process variations, varying resilience properties of different applications, and multiple reliable code versions. It thereby combines the novel contributions of the whole manuscript in a holistic optimization flow. In particular, the proposed *Dependability Tuning* system dynamically adapts the dependability mode at the hardware level through *resilience-driven redundant multithreading tuning* and at the software level through *variation-aware thread-to-core mapping*, and the *selection of reliable code versions* under given performance constraints, while considering the core-to-core frequency variations resulting from the manufacturing-induced process variations and run-time aging. Holistically considering knowledge from multiple system layers enables the proposed *Dependability Tuning* system to surpass state-of-the-art single-layer reliability optimizing solutions, as will be demonstrated using extensive comparison results in Chap. 7.

6.1   Reliability-Driven Offline System Software

Since the execution time of different function versions varies depending upon the input data, the *reliability-driven offline system software* determines schedule tables for all the functions such that these tables have appropriate function versions across all possible values of the execution time (i.e., remaining time to final deadline) and the reliability. These versions are selected while considering the already selected versions of the previous functions which collectively offers a reduced *Reliability-Timing Penalty* (*RTP*) when compared to other version combinations. At a certain time, selection of an appropriate version for a function to be executed next would depend upon the so far selected function versions, their execution time, reliability properties, and the application deadlines. To solve this problem, the proposed *reliability-driven offline system software* determines a *function schedule table* such that the resulting schedules minimize the overall *RTP* under different execution scenarios.

Figure 6.1 shows an abstract example of the schedule table for a function F_i where the x-axis and the y-axis show different values of the execution times and reliability, respectively. A single entry $G^*(i, r, t)$ inside the table stores the minimum *RTP* value of a certain version, and this value contributes to optimize the overall *RTP* (i.e., the case where the summation of the *RTP* values of all the function versions in a schedule gives the minimum *RTP* value). In the end, only the version indexes of the selected versions are stored in another table at an entry $j^*(i, r, t)$ which also corresponds to the reduced *RTP* value at an entry $G^*(i, r, t)$ in the previous table.

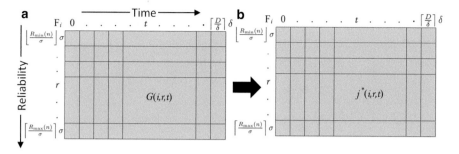

Fig. 6.1 Offline schedule table construction. (**a**) A table for function version F_i. (**b**) Version index of function F_i with the minimum expected RT penalty

Table 6.1 An example illustration for generating function schedules (selected function versions are highlighted in bold)

F_i	Version (j)	Reliability (RT) (R_{ij})	Execution time (t)
F_1	1	3	7
	2	**4**	**3**
F_2	**1**	**7**	**4**
	2	2	6
	3	1	5
F_3	1	3	2
	2	**5**	**1**

Therefore, only table $j*(i, r, t)$ is maintained that only has the version indexes to reduce the memory footprint. For instance, two bits may be required to store the indexes in contrast to storing the *RTP* values that may require more bits.

These tables are constructed using dynamic programming. In the following, an abstract example is presented to explain how these tables are constructed. Note, for simplification, in this example, the execution time is considered fixed (i.e., not varying which is typically not the case at run-time). Assume that an application is composed of three functions, i.e., F_1, F_2, and F_3, that have multiple versions as shown in Table 6.1, each with a certain value of R_{ij}, and the execution time t. The table construction starts from backwards, i.e., from the last function which in this example is function F_3 because if more reliable versions are selected for the previous functions that have longer execution times then this may lead to deadline misses, and thus should be avoided (see more details in Sect. 6.2).

The first step is to find the minimum and maximum range of the reliability values, i.e., R_{min} and R_{max}, for every function except F_1. In order to obtain this for the function F_3, it is required to add up the minimum and maximum reliability values found in the versions of previous functions, i.e., F_1 and F_2. Formally, the R_{min} values for the two-dimensional table for F_3 is: $R_{min} = \sum_{i=1}^{n} \min_j R_{ij}$, that leads to $R_{min} = 4$. Similarly, the $R_{max} = 11$ which is computed as: $R_{max} = \sum_{i=1}^{n} \max_j R_{ij}$. So the range of

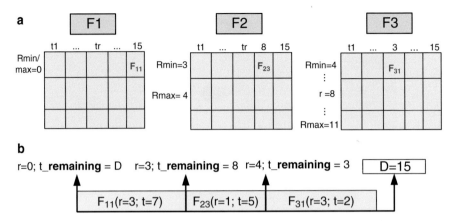

Fig. 6.2 (a) Design-time schedule tables, (b) run-time execution of the schedule

the reliability obtained is between R_{min} and R_{max}, i.e., 4, 5, ..., 11. Now this range means that, at run-time if the previously executed versions of F_1 and F_2 were the ones with the minimum R_{ij} values (i.e., version $F_{11} = 3$; $F_{23} = 1$), then the so far achieved reliability, before F_3 is executed, will be 4. Therefore, the reliability min/max range for the function F_3 will not go below 4, and above 11 (see Fig. 6.2a, case of F_3). In a similar way, the R_{min} and R_{max} for the function F_2 can be computed by looking at the minimum and maximum R_{ij} values in function F_1, which are 3 and 4, respectively (see Fig. 6.2a, case of F_2). However in case of F_1, no previous function is executed yet, therefore, the R_{min} and R_{max} both are equal to 0 which means that at the time when the execution starts for function F_1, the so far achieved reliability is 0 and the remaining time is the total time to the deadline. The reliability value denotes the so far achieved system reliability after these versions are executed.

The value of the execution time relates to the remaining time to the deadline. Prior to starting the execution, the elapsed time is 0 as no function has executed so far, which is the minimum time value on the table. However, the maximum time is the total time which is the final deadline D. The min/max range for the execution time of all the functions is the same, i.e., from 0 to D. Although the earlier functions are expected to finish before the deadline, at run-time the execution behavior might change and therefore version options for all possible values of the execution times are required. For F_1, the remaining time to the final deadline is the complete deadline time, in this case the best function version can be selected with minimum RTP. However, its execution time still needs to be considered because if the reliability-wise best version is selected that has a longer execution time, it may affect the version selection decisions for the subsequent functions. The probability distribution of the execution time of a function is taken into consideration so that the system can exploit the dynamic execution behavior by selecting suitable versions for the subsequent functions. Now given that the RTP values for all the function versions are precomputed, these RTP values are taken into account while deciding about an appropriate function version that should be placed inside a table entry across certain r and t values. Figure 6.2b denotes how at run-time these function versions are selected from the schedule tables for execution to compose a reliable application. $D-t$ denotes the remaining time until the deadline. Depending upon the current execution behavior of F_1 (i.e., its reliability penalty and the remaining time to the deadline), a suitable version for F_2 and similarly for F_3 is selected to dynamically link the corresponding binary codes of different functions and compose an application. In the following, the optimization objective and the *RTP* optimization for an application with one function and for multiple functions are explained.

6.1.1   Optimization Objective

The optimization objective is the *Reliability-Timing Penalty (RTP)* as explained in Sect. 4.4.2. Equation 4.14 is repeated here as Eq. 6.1 for the ease of discussion, where R denotes the reliability function and *miss rate* denotes the percentage of deadline misses for the application, and the parameter α is a user-defined weight to trade off the importance of functional reliability and timing reliability depending upon the system requirements.

$$RTP = \alpha R + (1 - \alpha) \, miss \, rate \qquad (6.1)$$

   The objective of the studied problem can be described as follows: Given an application with n sequential functions, the objective is to generate and execute the application software such that the overall *RTP* is minimized. If function F_i already misses the deadline, it is expected that the application will miss its deadline no matter how the rest of the functions are executed. However, it is the users' choice to further improve the reliability or further improve the performance. In case performance is preferred, the version with the minimum average execution time should be chosen. However, if the reliability improvement is preferred, the version with the minimum reliability penalty (hence, higher reliability) should be chosen. For the rest of the cases, it is assumed that θ_i is the preferred version of function F_i when F_i has already missed the deadline. For notational brevity, we define $\rho_i = \sum_{l=i}^{n} R_{l,\theta l}$ for the total reliability penalty from F_i to F_n when function F_i has already missed the deadline. Note, here the aim is to minimize the probability of program errors as a means of improving the overall system reliability. The proposed approach is orthogonal to other means like error detection and error recovery. Furthermore, a reduced probability to program errors also corresponds to a reduced number of recovery operations.

6.1.2   Optimizing for the Reliability-Timing Penalty

This section presents the system software level optimization, given multiple compiled function versions from Chap. 5, that aim at minimizing the total *RTP* for a set of functions. Let us start with the simplest case by considering applications with only one function and then moving further to consider multiple functions.

Case of One Function: When the application has only one function, the probability $\phi_{1,j}$ of deadline misses can be evaluated for a version j of function F_1, i.e., $\phi_{1,j} = 1 - C_{1,j}(D)$, where $C_{1,j}(D)$ is the probability that the execution time is less than or equal to the deadline D. Therefore, by selecting the version j, the *RTP* for the application with one function is defined as: $\alpha R_{1,j} + (1-\alpha)\phi_{1,j}$. It is clear that the version $j*$ that minimizes the RT penalty $\alpha R_{1,j*} + (1-\alpha)\phi_{1,j*}$ should be selected for execution.

Case of Multiple Functions: This subsection describes the minimization of the *RTP* when the application has multiple functions. The simplest way is to select the versions statically. However, finding the optimal selection with the minimum *RTP* statically is in general NP-hard.[1] The optimal static version selection with the minimum expected *RTP* may still be too pessimistic. Let's consider the following motivational example with two functions: Suppose that function F_1 is executed with one specific version that has high variability in the execution time, in which $R_{1,1} = 0.1$, $P_{1,1}(1) = 0.5$, and $P_{1,1}(9) = 0.5$. The function F_2 has two versions with fixed execution time, where $R_{2,1} = 0.1$, $P_{2,1}(9) = 1$, $R_{2,2} = 0.3$, and $P_{2,2}(1) = 1$. Suppose that the value of D is 10. There are only two options for the static version selection, i.e., with $\{F_{1,1}, F_{2,1}\}$ or $\{F_{1,1}, F_{2,2}\}$. For the case $\{F_{1,1}, F_{2,1}\}$, as the deadline miss rate is 50 %, the expected *RTP* is $0.2\alpha + 0.5(1-\alpha)$. For the case $\{F_{1,1}, F_{2,2}\}$, as the deadline miss rate is 0 %, the expected *RTP* is 0.4α. Since $0.4\alpha > 0.2\alpha + 0.5(1-\alpha)$ when $\alpha > 5/7$, for the above example, the following versions for function F_1 and F_2 should be selected. $\{F_{1,1}, F_{2,1}\}$ if $\alpha > 5/7$, and $\{F_{1,1}, F_{2,2}\}$, otherwise.

However, the above static assignment is pessimistic, as the function F_2 can react according to the varying execution behaviors of the function F_1. When the function F_1 finishes very early, i.e., at time 1, for function F_2 the high-reliability version $F_{2,1}$ can be adopted with longer execution time if the remaining time to the deadline is sufficiently large. In case function F_1 finishes very late (i.e., at time 9), then function F_2 has to run the low-reliability version $F_{2,2}$ with shorter execution time if the remaining time to the deadline is too small. The above dynamic version selection of function F_2 is with expected *RTP* equal to 0.35α, which is lower than the expected *RTP* 0.4α of the optimal static version selection $\{F_{1,1}, F_{2,2}\}$, when $\alpha \le 5/7$. Therefore, for the rest of this section, the dynamic version selection will be presented, in which a schedule table is prepared offline and the scheduler adopts the suitable versions of the functions according to the run-time execution behavior in an online fashion. Note that, here the timing behavior is analyzed based on the estimated probability functions to minimize the expected *RTP*.

The selection of the execution version for function F_i depends on (1) the execution behavior until now for the previously executed functions $F_1, F_2, ..., F_{i-1}$; and (2) the reaction for the future functions $F_{i+1}, F_{i+2}, ..., F_n$ according to the execution behavior of function $F_1, F_2, ..., F_i$. To capture the properties in the first part, it is imperative to know the total reliability penalty of functions $F_1, F_2, ..., F_{i-1}$ and how much execution time that the functions $F_1, F_2, ..., F_{i-1}$ have elapsed. All pos-

---

[1] It can be reduced from the multiple-choice knapsack problem.

sible scenarios for these properties are considered via exploring possible values. The properties in the second part will be captured by referring to a table entry where the reactions of $F_{i+1}, F_{i+2}, \ldots, F_n$ are stored. The proposed approach builds a 3D table $G( )$ for the execution behavior. Let $G(i, r, t)$ be an entry that stores the minimum expected *RTP* for the given n functions under the following conditions: $F_1, F_2, \ldots,$ F_{i-1} has finished at time "$D-t$," and $F_1, F_2, \ldots, F_{i-1}$ has total reliability penalty "r."

Furthermore, the decision of $G(i, r, t)$ also depends upon how the functions of $F_{i+1}, F_{i+2}, \ldots, F_n$ will react according to the execution behavior of function F_i. According to the above structure, $G(i+1, r', t')$ has to be built for all possible r and t values first such that the entries can be used when $G(i, r, t)$ is considered. Therefore, the procedure starts from the last function F_n. The entries for $F_{n-1}, F_{n-2}, \ldots, F_1$ are built later sequentially. To build $G(n, r, t)$, at first $j^*(n, r, t)$ is found which is the version of function F_n that has the minimum expected *RTP*, defined as follows; see Eq. 6.2.

$$j^* \left(n, r, t \right) = \begin{cases} \arg\min{}_j \; \alpha \left(r + R_{N,j} \right) + \left(1 - \alpha \right) \left(1 - C_{N,j} \left(t \right) \right), & t > 0 \\ \theta_n, & t \le 0 \end{cases} \tag{6.2}$$

Therefore, $G\left(n, r, t \right) = \alpha \left(r + R_{N,j^*} \right) + \left(1 - \alpha \right) \left(1 - C_{N,j^*} \left(t \right) \right)$, where j^* is $j^*(n, r, t)$. Now, let us consider the case to build an entry $G(i, r, t)$ where $i = n-1, n-2, \ldots, 1$. When $t \le 0$, $j^*(i, r, t)$ is θ_i and $G(n, r, t)$ is $\alpha(r + \rho(i)) + (1 - \alpha)$. Let us consider the other case when $t > 0$. If that function F_i selects version j, the following is known.

- The probability that function F_i finishes in execution time x (when $x \le t$) is $P_{i,j}(x)$
- The minimum expected *RTP* for the n functions has been calculated and stored in $G(i+1, r+R_{i,j}, t-x)$ when the execution time of F_i is x, where $x \le t$
- The probability when $x > t$ is $(1 - C_{i,j}(t))$ by executing all the functions $F_{i+1}, F_{i+2}, \ldots, F_n$ with their default versions $\theta_{i+1}, \theta_{i+2}, \ldots \theta_n$, respectively.

For notational brevity, let us define $H_j(i, r, t)$ as the expected penalty by using the above properties, where

$$H_j \left(i, r, t \right) = \left[\int_{x=0}^{t} P_{i,j} \left(x \right) . G\left(i+1, r + R_{i,j}, t - x \right) dx \right]$$
$$+ \left[\left(1 - C_{i,j} \left(t \right) \right) . \alpha \left(r + R_{i,j} + \rho \left(i+1 \right) \right) + \left(1 - \alpha \right) \right] \tag{6.3}$$

The first addend in the right hand side in Eq. 6.3 for the integration considers the convolution when the execution time of $F_{i,j}$ is no more than t, while the second addend considers the impact that $F_{i,j}$ already misses the deadline. Suppose that $j^*(i, r, t)$ is the index of j which minimizes $H_j(i, r, t)$ in the above equation. Therefore, for $i = n-1, n-2, \ldots, 1$, $G\left(i, r, t \right) = H_{j^*} \left(i, r, t \right)$, where j^* is $j^*(i, r, t)$.

An Example: Figure 6.3 gives an abstract idea of the design space of multiple versions of function F_1, F_2, and F_3 and the best selected versions during the table con-

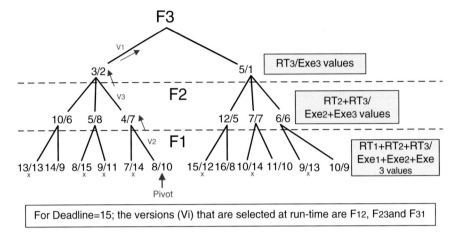

Fig. 6.3 Finding the best schedule after table construction at design-time

structions. This is a simplified example scenario to show how suitable function versions are selected which contribute to the minimum *RTP* of the complete schedule while also considering the *RTP* of the previously selected function versions. Note, for illustrative purposes, some example numbers of the *RTP* and the execution time are considered. It is shown that how multiple function versions are selected statically when the execution time is fixed. One potential way to solve this is using exhaustive search. However, to solve this problem with low complexity a dynamic programming algorithm can be employed to break the problem into subproblems, i.e., first finding the suitable version with reduced RT value for F_1, then for $F_3 + F_2$ and finally for $F_3 + F_2 + F_1$. The algorithm also stores the version of the function F_1 with the minimum expected *RTP* for each subproblem inside the table entry $j*(i, r, t)$. Finally, we backtrack the $j*$ table from the entry that achieves the minimum *RTP* for the complete problem in order to retrieve the version of each function. The initial pivot and the red arrows show one schedule comprising of versions F_{12}, F_{23}, and F_{31}, which can also be retrieved by backtracking. Now depending upon how the run-time execution scenario changes, e.g., due to the varying execution times of versions, the run-time system will choose suitable versions from the lookup table $j*$ that offers a reduced *RTP* while meeting the deadline. Note, in this work, soft real time applications are considered. This example of generating and executing the statically determined schedule is for the case when the execution times of the function versions are not changing (i.e., no variation in the execution time due to changing inputs). However, at run-time the execution time typically varies. Therefore, in order to consider these scenarios, the schedule tables are built offline using the dynamic programming algorithm for all possible combinations of execution time and reliability.

Complexity: Building the table $G(i, r, t)$ requires time complexity $O(K_i t) = O(K_i D)$. Therefore, by building $G(i, r, t)$ from $i = n$ to $i = 2$, $G(1, 0, D)$ can be built and $j*(1, 0, D)$ can be found, which provides the solution on how the first function F_1 should

be executed. Clearly, the function F_1 uses only one version $j*(1, 0, D)$. The other functions may require multiple versions in order to fully exploit the dynamic behavior of function executions. Let $R_{\max}$ be the maximum achievable reliability penalty of the system, i.e., $\sum_{i=1}^{n} \max_j R_{i,j}$. The above procedure requires time complexity $O(K_iD^2R_{\max})$ for building $G(i, r, t)$ for r in the range of 0 and $R_{\max}$ and t in the range of 0 and D. As a result, the total time complexity is: $O\left(\sum_{i=1}^{n} K_i D^2 R_{\max} \right)$. When all the possible values of execution time and reliability penalties are discretized, it is not difficult to see that the above procedure can optimally minimize the expected *RTP*. This can be proven by using the mathematical induction hypothesis, where the base case starts from function F_n. The above presentation requires to build the table for all possible values of t and r. However, it is not necessary to build some non-achievable entries in the table. For notational brevity, suppose that $R_{\max}(i)$ is:
$\sum_{l=1}^{i-1} \max_j R_{l,j}$ and $R_{\min}(i)$ is: $\sum_{l=1}^{i-1} \min_j R_{l,j}$. For building $G(i, r, t)$, only r in the range of $R_{\min}(i)$ and $R_{\max}(i)$ has to be considered. Moreover, the units of the time and the reliability penalties can be changed. For example, one can build the table based on the timing unit of 0.1 or 0.01 ms. The larger the timing unit and the reliability unit are adopted, the more the loss of the accuracy and the less the complexity for the table construction (see Algorithm C.8). The procedure is the same as the flow presented above. The time complexity of the offline table construction is
$O\left(\sum_{i=1}^{n} K_i \left(\frac{D}{\delta} \right)^2 \left(\frac{R_{\max}}{\sigma} \right) \right)$ and the space complexity is $O\left(n\frac{D}{\delta} \frac{R_{\max}}{\sigma} \right)$.

Summary: The schedule tables are prepared offline and an appropriate schedule is determined via backtracking. After the tables are constructed, the next step is to select an appropriate schedule at run-time that gives reduced *RTP*. Once this schedule is determined, the run-time system software will perform an application composition that selects different function versions from the schedule table, depending upon the current execution behavior (i.e., the reliability penalty and the remaining time to the deadline) and dynamically links the corresponding binary codes of different functions. Unlike the table construction procedure, the function executions start from the first function till the last. In the following sections, the function's execution scheduling for single core and multi-/manycore processors is presented.

6.2   Reliability-Driven Function Scheduling for Single Core Processors

Multiple functions are scheduled for execution in a sequential order on a single core processor architecture, such that the expected *RTP* for the overall system is minimized. The run-time system software performs version selection at run-time by

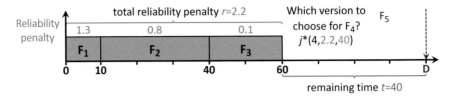

Fig. 6.4 RTP optimization and selection of multiple function versions inside the table

determining which function version F_i should be executed. Now the question is how the run-time system software finds a suitable F_i version for all the functions inside the offline generated tables. The first function version is executed by the version $j*(1, 0, D)$. But, the other functions F_i may have various versions, depending upon the so far achieved reliability penalty of functions $F_1, F_2, \ldots F_{i-1}$ and the remaining time to the deadline of this execution instance of the application. According to Algorithm C.8, if the remaining time to the relative deadline is t and the reliability penalty for the first $i-1$ functions is r, the run-time system software looks up the table entry $j*(i, r, t)$. However, when considering the timing unit δ and the unit σ of the reliability penalties, the run-time system software instead looks up the entries with $j*(i, \lfloor r/\sigma \rfloor \sigma, \lfloor t/\delta \rfloor \delta)$. Therefore, the binary version implementation $F_{i,j*}$ is selected, where $j*$ is $j*(i, \lfloor r/\sigma \rfloor \sigma, \lfloor t/\delta \rfloor \delta)$.

Figure 6.4 presents an example execution scenario that explains the run-time version selection procedure. Before the execution starts, the overall system's *RTP* is 0 and the remaining time to the final deadline is the complete execution time, i.e., D, since no function has been executed so far. For the first function F_1, there is only one entry in the table for the coordinates of reliability 0 and execution time D, therefore, there is only one function version option, i.e., $j*(1, 0, D)$, which gives minimum *RTP*. After the first function F_1 is executed, the run-time system software has to find an entry for the subsequent function versions. Let us suppose function F_2 and F_3 are also executed and the run-time system software is selecting a suitable function version for function F_4. After F_3 is executed, at 60 time units the run-time system software first analyzes the remaining time to the deadline which is 40 time units and the so far achieved total system reliability, which is the accumulated *RTP* of all the previously executed function versions $F_1, F_2,$ and F_3, i.e., $r=2.2$. The run-time system software will select an entry for function F_4, i.e., $j*(4, 2.2, 40)$, inside the table across for the coordinates $r=2.2$ (i.e., so far achieved system *RTP*) and $t=40$ execution time units (i.e., the remaining time to the final deadline D). The value for the *RTP* across the reliability 2.2 and execution time 40 is given in table $G(i, r, t)$, whereas the function version index is stored in $j*(i, r, t)$.

Once the table $j*$ is built, the entries of $j*$ should be stored in the main memory as a lookup table. With such a mechanism, deciding a version to be executed requires only $O(1)$ time. However, with this approach the space complexity becomes a problem. Moreover, this may result in many redundant entries. However, it is not necessary to keep all the entries of $j*$. For example, when $j*(i, r, t)$ remains the same within the range of $[r_1, r_2]$ and within the range of $[t_1, t_2]$, the run-time system software only needs to keep one entry for the index. Therefore, only the representative entries are stored. Moreover, some entries can be removed further if the difference of the *RTP* is too small between two entries in order to reduce the memory overhead. In general, the system designers can decide the tolerable overhead for the resulting table. Note that the table $j*$ is required to be protected so that the run-time system software can select the correct versions for minimizing the *RTP*. By removing the sparse entries, as discussed above, the memory protection overhead can be reduced.

Results and Discussion: Following the prominent industrial and research trends of AMD [63] and IBM [62], for the experiments, it is considered that the memory and caches are protected. The performance overhead is at most $(n-1)$ times the overhead of fetching one table entry from the main memory. This is considered negligible, compared to the execution time of one application iteration. Multiple function versions are profiled on the reliability-aware processor simulator using various different input data sets to obtain the distributions of errors, vulnerabilities, and execution time (see Sect. 5.4 for details on the generated function versions and their reliability/performance properties). The cumulative distribution function (CDF) of the execution time is generated and partitioned into ten different steps. The CDFs for different compiled versions of two applications are shown in Fig. 6.5.

For evaluation, various functions of different applications from MiBench are used, for instance, the *H.264* video encoder with three key functions *SAD, DCT,* and *SATD,* and other applications like *ADPCM, CRC, SusanS,* and *SHA.* In order to realize a complex real-world application scenario for reliable code generation and execution, all these applications were integrated into one application of "*secure video and audio processing.*" The following ordering of these seven functions is used: "*SAD, SATD, DCT, SusanS, CRC, ADPCM, and SHA.*" Figure 6.6 shows the results for the "*secure video and audio processing*" application simulated under three different settings of the fault rates, i.e., 10^{-6}, 10^{-7}, and 10^{-8}. According to the functions' *FVIs* and the fault rates, the corresponding reliability penalties are derived. When the relative deadline increases, the expected *RTP* decreases, since the timing constraint is less stringent. Therefore, when the relative deadline is too

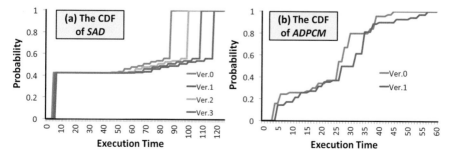

Fig. 6.5 The cumulative probability distribution (CDF) for two example applications from the MiBench suite

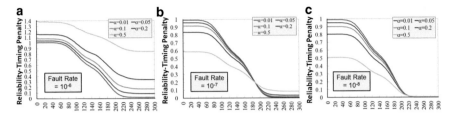

Fig. 6.6 Expected RTP under different fault rates. (**a**) 10^{-6} faults/cycle, (**b**) 10^{-7} faults/cycle, (**c**) 10^{-8} faults/cycle

small, i.e., less than 40 ms, the miss rate is almost 100 % for all the version selections, and, therefore, the most reliable versions will be selected (or the default versions will be enabled). On the other hand, when the relative deadline is large, i.e., more than 240 ms, the miss rate becomes nearly 0 %. A proper function version selection will have almost no deadline misses, and will try to minimize the reliability penalties as well. In all the three plots, the *RTP* values change more in the range when the relative deadline is between 40 and 240 ms, and the algorithm tries to balance the miss rate and the reliability penalties.

In Fig. 6.6, for one given α under a given deadline D, when the fault rate is high, it can be noticed that the expected *RTP* is also higher. This is because of the fact that R_{ij} is larger. When $\alpha=0.01$, the reliability penalties R_{ij} will not play a very significant role in our settings, while the deadline miss rate matters. As a result, similar trends for $\alpha=0.01$ can be observed in all the above subplots. In other words, the *RTP* trend for $\alpha=0.01$ is similar to the miss rate. When the fault rate is 10^{-6}, the setting with $\alpha=0.5$ is the one with the maximum expected *RTP* in the simulated cases, but it becomes the one with the minimum when the fault rate is 10^{-8}. The main reason comes from the settings of R_{ij}. When fault rate is 10^{-6}, the achievable reliability penalty is between 1.65 and 2.72. When fault rate is 10^{-8}, the achievable reliability penalty is between 0.0165 and 0.0272. Therefore, compared to the deadline miss rates, the reliability penalties play a significant role when the fault rate is 10^{-6}, a comparable role when the fault rate is 10^{-7}, and a very minor role when the fault rate is 10^{-8}. In Fig. 6.6a, even though the proposed run-time system tries to minimize the expected *RTP*, the values of αR_{ij} are still too significant, so that the *RTP* remains very high, especially when α is large. In Fig. 6.6b, the system tries to make a balance between the reliability penalties and the miss rate. However, since $\sum_i \min_j R_{i,j}$ is 0.165 (approximately) when the fault rate is 10^{-7}, the minimum expected RTP for $\alpha=0.5$ is still about 0.1.

In addition to the evaluation for the dynamic version selection, the evaluation for the system software for the following two static version selections is also presented:

- Min. R: chooses the version j with the minimum R_{ij} for each function F_i to improve the reliability.
- Min. Avg: chooses the version j with the minimum average execution time for each function F_i to improve the performance.

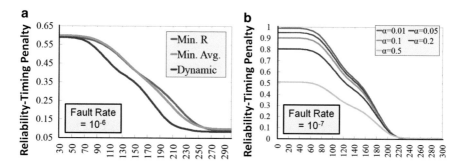

Fig. 6.7 Expected RTP under different fault rates by comparing to the static version selections when $\alpha=0.5$. (**a**) 10^{-6} faults/cycle, (**b**) 10^{-7} faults/cycle

Figure 6.7 presents the simulation results when the fault rate is 10^{-6} and 10^{-7} with $\alpha=0.5$ by considering the static version selections (Min. R and Min. Avg) and the dynamic version selections based on Algorithm C.8. Compared to the dynamic version selection, both of the static version selections are worse. When the fault rate is high (i.e., 10^{-6}), the static version selection with the minimum average execution time is worse, because the reliability penalties play an important role for reducing the RTP. This also explains why the static version with the minimum R_{ij} for each function F_i is better. When the fault rate is low (i.e., 10^{-7}), the difference between the two static version selections is very limited.

Summary: The functional and timing reliability is taken into account to improve the overall reliability at multiple system layers, i.e., offline system software and run-time system software. At offline, multiple schedule tables are built for different function versions, such that the expected RTP is minimized. At run-time, reliability-driven function scheduling is performed for a single core processor to minimize the overall RTP for the complete function schedule. The run-time system software selects suitable function versions for executions in a sequential fashion depending upon the so far achieved system reliability (due to previously executed functions) and the remaining time to the application deadline (considering the time elapsed by the execution of previous function versions). The contributions in this section targeted the soft error resilience for single core processors. However, the real-world multi-/manycore processors are also subjected to multiple reliability threats, i.e., soft errors, aging, and process variations. In such cases, the challenge is to achieve high soft error resilience (i.e., low RTP) while considering the process variation and aging-induced effects on the cores' frequencies. The contributions in the following Sect. 6.3 address this issue while accounting for the concept of multiple function versions and their varying reliability and execution time properties. In this way, Sect. 6.3 integrates the works from the Chaps. 4 and 5 in an integrated cross-layer reliability optimization flow, while also leveraging the analysis observations from Chap. 3.

6.3 Reliability-Driven System Software for Multi-/Manycores

6.3.1 Soft Error Resilience in the Presence of Process Variations and Aging Effects

In a real-world scenario, an on-chip manycore system is subjected to multiple reliability threats like soft errors, aging, and process variations. As discussed in Chap. 2, due to the design-time process variation and run-time aging, different cores on a chip or different chips on the same/different wafers run at different frequencies [19, 25]. The amount of aging (i.e., frequency degradation induced due to threshold voltage shift) varies depending upon the stress produced due to the workload and operating conditions [19, 117]. Since different cores in a manycore system are subjected to different amounts of stress as a result of varying workloads, the aging imbalance (i.e., frequency degradation of different cores at different rates) may further aggravate core-to-core frequency variations.

The current state-of-the-art in soft error resilience for multi-cores mainly employs *redundant multithreading* (*RMT*) [36, 67] that enables execution of redundant threads on different cores and performs the corresponding comparison and rollback/voting. However, such RMT techniques [36, 67] do not account for the process variations and aging effects during the soft error mitigation process, which may lead to *either* excessive mismatches/rollbacks because the result of the redundant thread is not available at the same time instant *or* significant synchronization delays that can violate the performance constraints (i.e., degraded timing reliability) of different critical tasks.

Figure 6.8 presents an example scenario illustrating the basic RMT process which is performed without any frequency variations. Three redundant tasks are executed on three different cores and all the three outputs are received at the same time *t1*. Afterwards, considering a TMR-based RMT, the voting is performed and the final output is received within the deadline (alternatively, a comparison and rollback is performed in a DMR-based RMT implementation). This represents the *nominal case* that may not exist in the current manycore processors due to the presence of core-to-core frequency variations. Figure 6.9 presents three different

Fig. 6.8 RMT without frequency variation

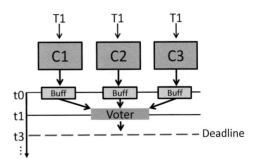

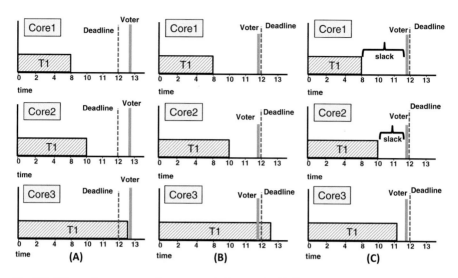

Fig. 6.9 Effects of core-to-core frequency variations on the RMT in timing-constrained scenarios (**A**) If voting is performed after the task finishes on the slowest core considering cores with different performance properties. (**B**) If voting is performed shortly before the deadline considering cores with different performance properties. (**C**) Potential solution where the allocation strategy takes the performance properties of the cores into account to select cores with similar properties

scenarios when the RMT is realized in the presence of core-to-core frequency variations, which may lead to unaligned arrivals of the outputs and output mismatch at the voter.

Scenario 1 in Fig. 6.9 scenario (a): Let us assume that we have three different cores (*Core1*, *Core2*, and *Core3*) exhibiting different performance properties and three redundant copies of the task *T1* are executing on these cores. If the voter is placed considering the performance properties of *Core3*, i.e., after all the task finishes their executions, this may lead to a deadline miss due to the synchronization issue and may ultimately lead to the timing errors because of the misaligned outputs of the cores.

Scenario 2 in Fig. 6.9 scenario (b): This scenario shows the case when the voter is placed near the deadline such that the voting is performed just before the deadline. However, this may lead to functionally Incorrect Outputs because *Core3* produces the output later than the deadline due to its slower performance. The delayed output in case of *Core3* may lead to timing errors. However, in case of *Core1* and *Core2*, a functionally Incorrect Output may be delivered because of soft errors since the remaining time before the voting is higher. The soft error rate on different cores can be different due to the process variations and aging effects. Now the question is: *how to perform the core allocation for the redundant tasks such that the outputs are received synchronously to avoid such issues while still ensuring both functional and timing reliability?*

Scenario 3 in Fig. 6.9 scenario (c): One potential solution to the above challenge could be to allocate the cores to the redundant tasks with similar/matching performance properties in order to reduce the performance gaps between the task execution properties and the cores' frequency. Such a solution is illustrated in this scenario, where *Core3* has a higher operating frequency (that fulfils the timing requirements of the redundant task of T1) compared to *Core3* of the scenario 1 and 2. Another interesting observation can be made when the tasks on *Core1* and *Core2* finish much earlier than the deadline resulting in the *slack.*[2] In such cases, a more soft error resilient compiled version can be selected for these redundant tasks, which may have a longer execution time but still fulfilling the deadline constraint, to further improve the soft error resilience. This would be beneficial especially in cases of DMR-based redundant multithreading because a more soft error resilient compiled version would have a reduced probability of errors and will consequently lead to a reduced number of rollbacks, thus a reduced performance overhead to ensure a reliable operation. The challenge here would be *to select an appropriate compiled version depending upon the core's frequency, deadline, and the execution properties of the selected version* that would result in a particular slack value.

6.3.2   *Dependability Tuning System for Soft Error Resilience under Variations*

In order to address the above-discussed challenges and to achieve efficient soft error resilience in manycore processors under process variation and aging-induced core-to-core frequency variations, this manuscript presents a novel *Dependability Tuning (dTune) system*; see Fig. 6.10. The *dTune* system leverages the knowledge of variable vulnerability and error masking properties of different applications and multiple reliable code versions, that provides important means to compensate for the process variation and aging-induced frequency variations, while at the same time providing distinct soft error resilience. Furthermore, our detailed error analysis and varying vulnerability/masking profiles in the earlier chapters illustrated that *not all applications may require the same level of dependability*. Furthermore, different compiled versions for different application functions exhibit distinct reliability and execution time properties. Hence the *diversity in the error distribution and masking properties of different application function versions is exploited for dependability tuning at the hardware and software levels, especially under resource competing scenarios.*

   The problem of compiler-driven dependability tuning poses the following challenges:

1. *Dynamically selecting an appropriate dependability mode* (e.g., activation or deactivation of redundant multithreading—RMT) in manycore processors for various applications under different performance and area constraints. Such an

---

[2] If the actual execution time is less than the estimated worst-case execution time, the unused time is called *slack*.

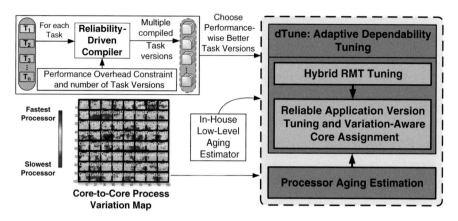

Fig. 6.10 The proposed dTune system for dependable application execution

RMT tuning (or in other words, RMT mode selection) needs to account for (a) diverse soft error vulnerability, error masking, and performance properties of different applications, and (b) the encountered error rates.

2. *Dynamically selecting an appropriate reliable code version* for each application considering the core-to-core frequency variations because of the design-time process variations and run-time aging-induced performance degradation.

3. *Mapping the selected version on the allocated set of cores at run-time* such that, in case of RMT, the execution properties of the redundant threads are closely matched considering the core-to-core frequency variations in the underlying hardware.

To evaluate such a dependability tuning system, there is also a need for a *processor aging estimation framework* (see details in Chap. 7).

Figure 6.11 presents the detailed operational flow and overview of the proposed *dTune* system. The inputs to the *dTune* system are: process variation map and aging estimates for all cores under a given run-time scenario, a set of concurrently executing tasks along with their multiple reliable code versions, and their execution time and reliability properties. Our *dTune* system improves the system dependability through the following two key operations (explained in detail in the subsequent sections): (1) Dynamic RMT Adaptations and Core Allocation, and (2) Dynamic Reliable Code Version Selection.

Formal Notations: Before proceeding to the algorithms of different components of the dTune system, the formal notations for hardware and application are presented.

Hardware Architecture: We consider a manycore processor $C = \{C_1, C_2, \dots C_N\}$ with N ISA-compatible homogenous RISC cores (e.g., a pipelined LEON3 embedded processor). These cores are heterogeneous w.r.t. their performance capabilities due to the design-time process variations or run-time NBTI aging effects. Each core has a private instruction and data cache. Due to the varying workloads, different cores

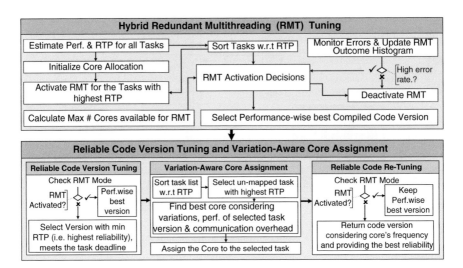

Fig. 6.11 Operational flow of the proposed dTune system

execute at different frequencies to ensure correct execution and to avoid delay faults, i.e., each C_i is associated with a frequency f_i. In this work, RMT can be activated in the DMR or TMR modes.

Application: An application is given as a task graph $G = (T, E)$, where T is a set of M nodes representing tasks, such that $T = \{t_1, t_2, \ldots, t_M\}$. E is the set of edges representing task dependencies: $E = \{e_{xy} \mid (t_x, t_y) \in T\}$. Each task t_j is compiled using the reliability-driven compiler that generates K_j task versions: $t_i = \{t_{(i,1)}, t_{(i,2)}, \ldots, t_{(i,Kj)}\}$. Each $t_{(i,k)}$ version has distinct reliability and performance properties (in terms of execution time) $L_{(i,k)}$. The task reliability is quantified using the *Reliability-Timing Penalty* (*RTP*) metric that jointly takes the functional and timing reliability into account. The functional reliability is quantified as the *Function Vulnerability Index* (*FVI*). For a given version, the timing reliability is given as the probability of deadline miss ($\phi_{(i,k)}$), e.g., due to a slow core as a result of process variation and/or aging. For a given raw error rate ER_{raw}, the formula of *RTP* in Eq. 6.1 can be revised as Eq. 6.4 in order to obtain the net error probability for a given function version. A small value of *RTP* corresponds to a more reliable task execution.

$$RTP\left(t_{(i,k)}\right) = \left(\alpha \times \left(FVI_{(i,k)} \times ER_{raw} \times L_{(i,k)}\right)\right) + \left((1-\alpha) \times \phi_{(i,k)}\right) \qquad (6.4)$$

6.3.3   Dynamic RMT Adaptations and Core Allocation

To enable hardware-level dependability tuning, the concept of *hybrid RMT tuning* is employed. For different concurrently executing application tasks, the RMT activation/deactivation decisions and variation-aware core allocation to all the redundant tasks are performed at run-time. It jointly accounts for the applications' reliability and performance properties along with the history of encountered errors at run-time under given area constraints and task deadlines. An appropriate RMT mode is selected at run-time depending upon the *RTP* and the performance requirements of the currently executing tasks, available number of cores, and the history of encountered errors. The RMT is activated for the reliability-wise critical tasks (i.e., tasks with the highest *RTP*) while considering their deadlines. Figure 6.11 illustrates the key steps for hybrid RMT tuning. Detailed algorithms are provided in Appendix C to ensure reproducibility.

Hybrid RMT Tuning: It jointly accounts for application-level knowledge of *RTP* and the hardware-level monitored error history. The goal is to assign an appropriate RMT mode to the tasks based on their *RTP*, performance properties, and performance variations in the underlying hardware. At first, the tasks' *RTP* and performance are estimated, a task list is created which is sorted by the *RTP*, and the number of free cores is calculated. In order to adapt the RMT mode, a history of outcomes of RMT, i.e., *error* or *no error*, is maintained after the comparison of redundant threads. Since the task which has the highest *RTP* is the reliability-wise most critical task, it is always executed in the RMT mode. In case errors are encountered, the history is updated and the RMT mode of other tasks is adapted. The underlying rational is: in case of intermittent or transient faults, other tasks may also get affected. Since the critical task has the highest vulnerability, it is more likely for these instructions to experience errors during their executions (this may change depending upon the fault location). However, if the vulnerability of another task is increasing, its *RTP* will also increase and this task will be set to execute in the RMT mode. In order to facilitate baseline execution for each task, at least one core is allocated to every task. In case of resource scarcity, the RMT mode is turned-off selectively, and the tasks with RMT are downgraded to free the cores for other demanding tasks in the priority order determined by *RTP*. Afterwards, the remaining cores are allocated to the other reliability-wise important tasks (i.e., with respect to their *RTP* values). In this case, the history of encountered errors is evaluated in a weighted manner giving a higher importance to recently conducted comparisons. Note, in case of dependent tasks where the previous output gives a mismatch, it is assumed that the error is recovered by task re-execution before its output is served as an input to its dependent task.

6.3.4   Dynamic Reliable Code Version Selection

To enable software-level dependability tuning, a *dynamic reliable code version selection* technique is employed. It selects an appropriate reliable code version for various tasks and their redundant task copies considering the selected RMT mode. Afterwards, an appropriate core, from the allocated set of cores to a task, is assigned for its execution. The core assignment takes core-to-core frequency variations into account such that the performance capabilities of the assigned core match the average-case execution time properties of the selected code version. A suitable core with distinct attributes (i.e., frequency) is taken into consideration while activating the redundant execution. For instance, if a task has a high *RTP*, a slower core might not be a suitable option as it may increase the number of program errors during single execution or it may increase the number of mismatches in case of RMT. Similarly, a relatively less-aged core may be beneficial to be assigned to a task with high *RTP* while relatively more-aged cores with performance degradation may be allocated to tasks with low *RTP*. Once the *variation-aware core assignment* is performed, the performance capabilities of the assigned core are known. Therefore, the *reliable code version can be re-tuned*, i.e., a more reliable code version can be executed within the time budgets due to the available *slack* while still meeting the task deadline. Figure 6.11 illustrates key steps for reliable code version tuning and the variation-aware core assignment. Details of methods adopted for these steps are explained below and the corresponding algorithms are provided in Appendix C.

The reliable code version tuning is performed in the following three steps (see details below). First, the initial version selection is performed based on the task deadline and *RTP* properties. Afterwards, a suitable core from the task's allocated pool of cores (considering process variation and aging-induced effects) is assigned to this task version such that the performance capabilities of the assigned core match with the execution time properties of the task version. In the last step, as the core properties are known, the version re-tuning is performed, i.e., the version selection is refined to match the assigned core's performance capabilities as much as possible in order to compensate for the aging-induced effects. Note that such a selection cannot be performed at design-time because the amount of frequency variation due to aging and the decision of a suitable task to core mapping cannot be predicted at design-/compile-time.

Step 1—Reliable Code Version Tuning: In this step, an appropriate soft error resilient compiled version is determined depending upon the task deadline and its *RTP*. It differentiates between the tasks *with* and *without* RMT mode assigned. For tasks with RMT (i.e., having more than two cores supporting TMR-based RMT), the performance-wise best version is selected as faults can be detected and corrected through redundant execution and voting, thus meeting deadline is the utmost priority. In other cases, the reliable code version with the lowest *RTP* (i.e., best reliability) is selected that can meet the deadline constraint when running on the slowest free core as the initial step and then refines in the following Step 3 considering frequency variations of different allocated cores.

Step 2—Variation-Aware Core Assignment: It accounts for process variations and aging-induced frequency degradation by assigning an appropriate core (i.e., with similar performance properties) to given tasks and their redundant copies. The assignment begins with the reliability-wise critical task first, i.e., tasks are sorted w.r.t. their *RTP*. It is beneficial for tasks with the highest reliability and deadline demands to obtain the fastest cores. For the RMT case, three performance-wise best free cores are evaluated for matching the task execution time (considering its communication time and performance on the core) and the core offering the best solution is selected. The aim is to balance the communication overhead and the performance that the particular core provides. In case a task executes in the RMT mode, the cores selected to execute the redundant threads of the task are the ones having the smallest distance to the core already selected. This is important for avoiding any unnecessary performance penalty while comparing the results of the redundant threads. For example, in case of RMT, if the core(s) executing the redundant thread(s) is/are located farther than the core executing the original thread, it may happen that the communication overhead of a farther core is too high and may cause a delay in determining the final result. In this case, a relatively slower core but near to the core with the original thread might be a better option in order to avoid the communication overhead. However, if a farther core has the best speed and the task version running on it is also the best from the performance perspective, then it may happen that the output is delivered well in time. In this case, it might be beneficial to select the best core which may be farther away.

Step 3—Reliable Code Version Re-Tuning: In the final step, the potential for task version improvements based on the individual performance characteristics of the selected core is exploited. For tasks running in the RMT mode, the performance is balanced based on the slowest core, i.e., selecting a reliability-wise better version on a faster core which can still finish its execution at the same time as the task on the slowest core running the fastest available task version. Otherwise, it is analyzed if the task version can be upgraded from the reliability perspective while meeting the deadline, as the actually selected core is now known.

An Example: Figure 6.12 shows an example illustrating the basic procedure of the *dTune* system highlighting the hardware-level and software-level dependability tuning. At first, the cores with similar performance properties are selected to minimize the performance gaps. However, still there are some available slacks which are exploited to select an appropriate task version that finishes before the deadline. The two major steps of the *dTune* system are shown in the figure: (1) Cores allocated with approximately similar performance properties, (2) Exploiting the slacks to execute more reliable compiled task versions in RMT which will lead to an aligned output at the voting before the deadline, thus minimizes the output mismatches.

Summary: In this section, an adaptive *Dependability Tuning (dTune) system* for manycore processors is presented that leverages the knowledge of varying *RTP* and execution time properties of different compiled versions of different tasks along with the core-to-core frequency variations (resulting from the manufacturing-induced process variations and run-time aging). The *dTune* system dynamically adapts the dependability mode at the hardware level through hybrid redundant multithreading tuning and at the software level through selection of reliable code version under given performance constraints.

Since the *dTune* system combines all the novel techniques proposed in this manuscript to realize a cross-layer reliability optimization flow, the detailed results and experimental evaluation and comparison to different state-of-the-art single-layer techniques under constrained scenarios is presented in Chap. 7. Furthermore, the experimental setup and implementation details along with different inputs are provided in Appendix A.

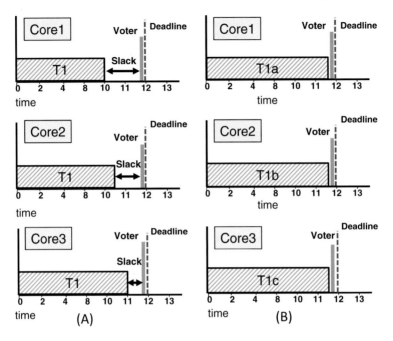

Fig. 6.12 Exploit slack and leveraging multiple reliability-driven compiled task versions. (**A**) Execution on cores with different performance properties leads to different slacks, (**B**) Exploiting the slacks to execute more reliable task versions while still meeting the deadline

6.4   Chapter Summary

This chapter presented three key contributions that leverage multiple system layers (i.e., compiler, offline and online system software, and partly hardware to obtain monitored statistics) to enhance the soft error resiliency in single core processors and in multi-/manycore processors subjected to process variations and aging. An offline system software generates multiple function scheduling tables considering the varying vulnerability and execution time properties of different compiled versions of different functions/tasks in an application while minimizing the Reliability-Timing Penalty. Afterwards, a run-time system for single core processors is proposed that selects appropriate function versions for dependable application composition in order to improve the overall soft error resilience. It accounts for the so far achieved reliability and the remaining time until the deadline depending upon the previously executed functions. Experimental results demonstrate that the proposed techniques balance the miss rate and the reliability penalties. However, this technique cannot be employed in multi-/manycore systems as the overhead of the schedule table for each core will be very high. Also, in a manycore system, other reliability threats (like process variations and aging) need to be considered during the soft error resilience, because these threats manifest as core-to-core frequency variations that will modify the execution time properties of different functions. Furthermore, a manycore processor may feature architectural support for redundant multithreading that is not exploited by this run-time system for the single core processors.

To address these limitations, a novel Dependability Tuning system is proposed for multi-/manycore processors that dynamically adapts the dependability mode (1) at the hardware level by selectively activating and deactivating the redundant multithreading mode for different concurrently executing applications; and (2) at the software level by selecting an appropriate compiled version for each task and its redundant thread, and mapping these compiled versions to appropriate cores that match the performance requirements of the threads. The proposed Dependability Tuning system aims at improving the soft error resilience by minimizing the accumulated Reliability-Timing Penalty for all the concurrently executing applications under core-to-core frequency variations.

The proposed contributions surpass the state-of-the-art redundant multithreading techniques by holistically accounting for the multiple reliability threats during soft error resilience, interactions between compiler and system software through exchange of multiple reliability versions, and joint consideration of functional and timing reliability. Since the Dependability Tuning system integrates the novel techniques of this manuscript in an integrated optimization flow, the detailed evaluation and comparison to state-of-the-art is provided in Chap. 7 that will demonstrate the benefits of employing a cross-layer software reliability optimization flow compared to single-layer solutions. The detailed experimental setup is explained in Appendix A.

Chapter 7
Results and Discussion

This chapter presents reliability improvement results of the proposed cross-layer reliability optimization flow (integrating all the novel contributions of this manuscript) compared to state-of-the-art single-layer reliability optimizing techniques. Evaluation of individual contributions compared to their relevant state-of-the-art techniques has already been presented in the respective chapters, i.e., Chaps. 4 and 5. First the processor synthesis, aging estimation, and process variation maps are presented in Sect. 7.1. Section 7.2 presents different benchmark applications from the MiBench benchmark suite [111]. Section 7.3 presents an overview of comparison partners, parameters considered for evaluation, and an overview of the results explained in subsequent sections. Section 7.4 presents the summary of comparison results for various chip sizes, numerous process variation maps, and various scenarios of simultaneously executing applications. Sections 7.5 and 7.6 present more in-depth results for different chip sizes, different application scenarios, and selected chips, respectively.

7.1 Processor Synthesis and Performance Variation Estimation

7.1.1 Processor Synthesis

To analyze the reliability of a circuit, the target design needs to be synthesized first to obtain the gate-level netlist where the characteristics of individual gates originating from a technology library can be considered afterwards. The *Synopsys Design Compiler* [122, 123] is used to synthesize the LEON3 processor and to obtain a netlist file. The Synopsys Design Compiler is a logic synthesis tool and is part of the Synopsys EDA tool chain. Prior to logic synthesis, the behavior-level target design needs to be verified in order to examine if the design is fulfilling the requirements. For that the *ModelSim*

© Springer International Publishing Switzerland 2016
S. Rehman et al., *Reliable Software for Unreliable Hardware*,
DOI 10.1007/978-3-319-25772-3_7

Table 7.1 Processor synthesis results for area and power

| Component | Area [#gate equivalents] | Component | Power [mW] | |
			Leakage	Dynamic
Pipeline	7.77E+03	Pipeline	7.34E−02	7.88E−01
Cache	4.40E+03	Cache	3.89E+00	7.48E+01
Register	2.87E+03	Register	2.60E−01	4.29E+00
Others	4.38E+03	Others	1.01E−01	1.14E+00
Total	19.42E+03	*Total*	8.53E+01	

[124] simulation tool is used. The LEON3 processor is synthesized with the following constraints: system clock frequency=300 MHz, 0.81 V, and junction temperature=125 °C. The input to the Design Compiler are the VHDL files of the target design, a TSMC 45 nm low-power standard cell library file and an execution TCL script. Depending upon the technology library, the characteristics and process corners vary, e.g., gate length of the transistors, operating temperatures and voltages, and timing properties. For the Instruction Cache (I-Cache) and Data Cache (D-Cache), the configuration setting is 1 set, 4 Kbyte/set and 32 Byte/line. For the Memory Management Unit (MMU), the configuration setting is eight instruction TLB entries, eight data TLB entries, fast write buffer, and 4k MMU page size. The output of the Design Compiler is the gate-level netlist file of the target design, standard delay files, and a report of area, power, and timing estimation result for the target design. The area and power results are shown in Table 7.1. *ModelSim* [124] is used for gate-level simulations and to generate the signal probabilities that are later used for aging estimation.

7.1.2 Processor Aging Estimation

The processor aging tool chain, developed in this manuscript, estimates the aging of a processor's critical path under varying activities/workloads. The inputs to the developed processor aging estimator are the netlist and aging estimates of various logic elements obtained from an in-house low-level aging estimator (developed together with the Virtherm-3D team from SPP1500 [118]) that considers the NBTI-induced aging. As discussed in Sect. 2.2, NBTI-aging manifests as the increase in the threshold voltage V_{th} by an amount ΔV_{th}. To compensate this effect, the circuit is required to execute at a lower frequency by a factor of Δf, otherwise the circuit output may be faulty due to the timing errors. The device-level NBTI aging model (Eq. 7.1) is based on the reaction–diffusion theory [13, 40].

$$\Delta V_{th} = 0.05 \times e^{-1500/T} \times V_{dd}^4 \times y^{1/6} \times d^{1/6} \tag{7.1}$$

ΔV_{th} is the mean threshold voltage shift in Volts, T is the temperature in Kelvin, V_{dd} is the supply voltage in Volts, y is the age of the transistor in years, and d is the duty cycle (i.e., probability that the transistor is stressed). The aging estimator for logic elements is based on ngspice-23 and requires device parameters such as transistor dimensions, activity, temperature, and load capacitance. The low-level aging results for an example NOR gate for different signal probabilities are shown in Fig. 7.1b.

In order to obtain the frequency degradation of a processor based on the delay degradation values of different logic elements (like SRAM cells, latches, and gates), in the first step its synthesized netlist is parsed and transformed into a graph-based data structure that simplifies traversing. Afterwards, a critical path analysis is performed and the delay degradation of the top $x\%$ critical paths of the given processor are estimated over 10 years, where x denotes a user-provided parameter. Note that the consideration of the temperature effects of thermal management and thermal-aware processing is beyond the scope of this manuscript and therefore the worst-case temperature (i.e., 125 °C in this case) is taken into account here. The switching activities/signal probabilities are obtained through *ModelSim* simulations. The initial delay estimates are obtained from the *Standard Delay Files*, which are obtained as an output of the logical synthesis. The developed tools generate aging estimates for the top $x\%$ critical paths, by applying the delay degradation to individual logic elements and re-performing the critical path analysis. Finally, the aging of the processor's critical path is estimated as the accumulated delay of all logic elements considering their respective duty cycles over several years (see Eq. 7.2):

$$\Delta \mathrm{Delay}\left(\mathrm{CP}\right) = \sum_{\forall \mathrm{le} \in \mathrm{CP}} \left(\mathrm{Delay}\left(\mathrm{le}\right) + \Delta \mathrm{Delay}\left(\mathrm{le}, d, y\right)\right) \qquad (7.2)$$

where Delay(le) is the un-aged delay of the logic element (in particular gates) while ΔDelay(le,d,y) is the delay degradation which is proportional to a logic elements' ΔV_{th}. The aging results after analyzing a LEON3 processor are shown in Fig. 7.1a. The highest increase of the delay of the critical path is observed in the first year, while over the period of 10 years an aging factor of more than 30 % is possible.

7.1.3 Process Variation Maps

Several chip process variation maps were generated for the experiments using the methodology and model explained in Sect. 2.3. These variation maps serve as an input to the reliability-aware manycore simulator, where each process variation map

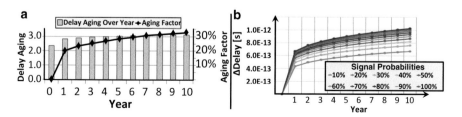

Fig. 7.1 (**a**) Processor aging in terms of critical path delay degradation (absolute and percentage degradation values) for different years considering 125 °C. (**b**) Aging of a NOR Gate at 125 °C

is represented as an individual file containing a grid with $Y \times Y$ grid points that define the frequency variation at its specific position, where Y is a flexible parameter determining the granularity of the grid (in this work $Y=80$). Each variation map is overlayed with the layout of the processor architecture that is planned to be analyzed in the following simulations allowing to investigate the impact of intra- and inter-chip frequency variations. Section 7.6 presents the core-to-core frequency distribution for two example variation maps (out of 100 different maps used in the experiments) when being overlayed with a 6×6 core processor consisting of homogeneous LEON3 cores. To determine the frequency of each individual core in the first step, the position and dimension of each individual core of the processor have to be defined. Afterwards, the grid is overlayed and for each core the grid points being mapped to it are determined, and their minimum is calculated that defines the frequency that can be assigned to this specific core.

7.2   Benchmark Applications

For evaluation, various applications from MiBench [111] (i.e., *ADPCM, CRC, SHA,* and *SusanC*) and an entire *H.264* video encoder [130] that exhibits various compute-intensive functional blocks (e.g., *SAD, SATD, DCT, HT, MC-FIR, IPRED*) with diverse computational properties are used. Several mixes of these applications are generated to realize competing scenarios of concurrently executing applications like a real-world scenario of secure video conferencing and random mixes. For each application, various thread instances are created that process a completely different set of data. Note that the selected applications have significantly varying vulnerability and error masking properties.

7.3   Comparison Partners and Evaluation Parameters

The subsequent sections present the evaluation and results for the proposed *adaptive Dependability Tuning (dTune) system* for manycore processors, which employs a cross-layer reliability optimization flow. In particular the results will illustrate the benefits of a cross-layer software reliability optimization (i.e., *dTune*) over software program-level and architecture-level single-layer soft error resilience techniques. The reliability evaluations will be performed for manycore processors of different sizes (i.e., number of cores) considering core-to-core frequency variations due to manufacturing variability and aging-induced effects. To realize reliability optimization under resource- and timing-constrained scenarios, evaluations are done for a large set of application mixes such that providing Redundant Multithreading (RMT) in the TMR mode for all the concurrently executing applications is not feasible for the given chip sizes. In such cases, it is more beneficial to provide RMT for the less-resilient applications while executing more-resilient applications without RMT.

In the following, a short overview of different comparison partners is presented, followed by a brief discussion on different parameters used for evaluations and highlights on the comparison results that will be discussed in the subsequent sections.

7.3.1 Comparison Partners

The proposed *dTune* system integrates the novel reliability optimizing techniques at different system levels as proposed in this manuscript. In summary, it employs the following information and methods at different system levels to realize cross-layer reliability optimization.

- *Hardware-Level*: process variation map, aged core frequencies at different years, microarchitecture-level analysis of spatial and temporal vulnerabilities for different instructions to obtain function vulnerabilities, area and fault probabilities of different processor components, and architectural support for redundant multithreading (RMT).
- *Compiler-Level*: multiple compiled versions for each task with varying reliability (quantified as RTP) and performance properties (quantified as execution time).
- *System Software-Level*: resilience-driven core allocation for RMT, mapping of redundant tasks on different cores considering core-to-core frequency variations, selection of appropriate code versions.
- *Application-Level*: deadline and workload information of different applications.

The reliability-timing results of the *dTune* system are compared to the following state-of-the-art techniques.

1. **CRT—Chip-Level Redundant Threading Technique** [36]: This technique targets maximizing the functional reliability by activating RMT in the TMR mode for all tasks while the tasks with the earliest deadline get the fastest cores. However, if all cores are utilized upcoming tasks have to wait for other tasks to finish before they can start executing their redundant threads.
2. **RTO—Reliability-Timing Optimizing Technique** [141]+[142]: This technique jointly optimizes for functional reliability (in terms of vulnerability) and timing (performance) considering individual applications [141] extended with the proposed concepts of different versions [142] to realize a more fair comparison.
3. **TO—Timing Optimizing Technique**: This approach aims at minimizing the probability of deadline misses by assigning the fastest core to the task with the earliest deadline. In order to avoid deadline misses it does not activate RMT for the tasks and preserves free cores for the upcoming tasks.

The *TO* is a performance optimizing solution while *RTO* and *CRT* are single-layer reliability optimizing solutions. *RTO* is program-level soft error resilience technique that uses reliable code versions, i.e., leveraging the compiler-layer for reliability optimization. *CRT* is an architecture-level soft error resilience technique that employs redundant multithreading in the TMR mode, leveraging the architecture-layer for reliability optimization.

Beyond the Comparison with CRT: Note, there exists similar concepts to *CRT* like Process-Level Redundancy (*PLR*) [37] and *Reunion* [38]. *PLR* is a system software-level technique for redundant multithreading that employs data structures in the memory managed by the operating system, instead of special voting hardware/cores and data buffers. In general, comparison of our *dTune* system to *CRT* would reflect potential reliability benefits compared to *PLR* and *Reunion* techniques due to the conceptual similarities between *CRT*, *PLR*, and *Reunion*. However, since developing operating system features is out of the scope of this manuscript, we rely on architecture-level support for redundant multithreading which is easier to implement in the simulation environment at the core granularity. Therefore, detailed comparison results are only provided for the *CRT* [36] technique.

Extra Information Provided to State-of-the-Art for Fairness of Comparison: The abovementioned state-of-the-art techniques do not account for process variation and aging during their soft error optimization process. However, for a fair comparison, these techniques are provided with the information regarding the cores' frequencies and compiled application binaries.

Evaluation Metric: As an evaluation metric, we employ the *Reliability Profit Function* (*RPF*) with *TO* as a baseline for comparison. It demonstrates the collective reliability savings in the form of *Reliability-Timing Penalty* (*RTP*) improvements for all application tasks $t \in T$. Consideration of *RTP* accounts for both functional reliability (i.e., *FVI* as the task's vulnerability to soft errors) and timing reliability (as the probability of deadline misses). Note, a high *RPF* value is reliability-wise better.

$$\mathrm{RPF} = 100 \times \left(1 - \frac{\sum\limits_{\forall t \in T} \mathrm{RTP}(t)_Z}{\sum\limits_{\forall t \in T} \mathrm{RTP}(t)_{\mathrm{TO}}} \right); \quad Z \in \{\mathrm{RTO,,CRT,,dTune}\} \qquad (7.3)$$

7.3.2   Parameters Considered for the Evaluation

For comprehensive evaluation, *RPF* savings of *dTune* are compared to different techniques for different chip sizes, numerous process variation maps, different aging years, and various number of application mixes that will demonstrate the range of reliability savings of the proposed cross-layer approach compared to single-layer techniques. In particular, different values of the following parameters are considered for a wide-coverage and comprehensive evaluation.

1. **Different Chip Sizes**: In order to account for the different chip sizes (in terms of number of cores) of a manycore processor, different architectures are analyzed ranging from a 6×6 core processor to a 12×12 core processor.

2. **Chip Maps with Process Variations**: Due to the design-time process variations, cores of different chips have different performance characteristics (in terms of their operating frequency) even for the same architecture, i.e., *chip-to-chip frequency variations*. Additionally, the performance characteristics of cores on the same chip vary, i.e., *core-to-core frequency variations*. In the experiments, 100 different process variation maps are evaluated for each chip size, where the frequency of each individual core on the chip is extracted for every map. This in practice would correspond to an evaluation for 100 different processor chips of a given size (i.e., 6×6, 8×8, 10×10, or 12×12).
3. **Scenarios of Application Mixes**: To account for different workload scenarios and to consider dependent application sequences with distinct reliability, execution time and deadline requirement characteristics, different scenarios with diverse application mixes are generated and used for evaluation. Application mixes differ in terms of concurrently executing application tasks, their execution sequence and interdependencies. Among these mixes, many represent real world use cases, for instance, secure video conferences. Besides this, several random mixes of applications are considered. These mixes exhibit different number of parallel executing applications that require a varying number of cores for concurrent execution (e.g., 5–55 application scenarios being executed in parallel on an 8×8 core architecture).
4. **Aging Years**: For showing the influence of run-time aging on the soft error resilience decisions of the proposed *dTune* system, RPF results are compared for the initial state of the processor and after 1, 5, and 10 years of aging.

7.3.3 An Overview of the Comparison Results

In the subsequent sections, following comparison results will be presented at different abstractions for detailed discussion. Starting with the summary results in Sect. 7.4, more details are presented in Sects. 7.5 and 7.6.

1. *Analyzing Overall Savings for Different Aging Years*: A box plot showing the overall summary of the *Reliability Profit Function* (*RPF*) of *RTO*, *CRT*, and *dTune* compared to *TO* summarizing all cases of different chip sizes, scenarios of application mixes, and process variation maps over different aging years.
2. *Analyzing Overall Savings for Different Chip Sizes and Aging Years*: Individual box plots show the *RPF* improvements of different comparison partners for different chip sizes and different aging years, summarizing all cases of application mixes and process variation maps.
3. *Analyzing Detailed Savings for Different Application Mixes, Chip Sizes, and Aging Years*: Detailed bar graph plots for different chip sizes and different aging years, where each bar graph plot illustrates the *RPF* improvements of different comparison partners for different application mixes with different number of parallel executing applications. Each bar in the plot represents an averaged *RPF* value over all process variation maps.

Fig. 7.2 An abstract
illustration of the box plot

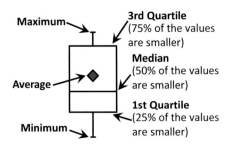

4. *Analyzing Detailed RPF Savings for Two Chips with Different Core-to-Core Frequency Distributions*: Detailed RPF savings are shown in form of bar graphs for two selected process variation maps for a 6×6 processor chip. Moreover, core-to-core frequency distribution is also shown for different aging years to relate variations in the *RPF* savings to the frequency variations and thus highlighting the available optimization potential.

Box Plots: A box plot corresponds to a distribution of results and represents the following five important evaluations as shown in Fig. 7.2. (1) *Minimum* corresponds to the 0 % Quartile, i.e., the minimum value of the complete set of results. (2) *Maximum* corresponds to the 100 % Quartile, i.e., the maximum value of the complete set of results. (3) The *First Quartile* corresponds to the result for which 25 % data points in the complete result set are equal to or smaller than this result value. (4) The *Second Quartile* corresponds to the result for which 50 % data points in the complete result set are equal to or smaller than this result value. This is also the median value of the complete data set. (5) The *Third Quartile* corresponds to the result for which 75 % data points in the complete result set are equal to or smaller than this result value. Additionally, the *Average Value* is overlayed with the box plot in form of the diamond symbol.

7.4 Overview of Savings Compared to State-of-the-Art

Figure 7.3 shows the box plots illustrating the overall *RPF* results of *dTune*, *RTO*, and *CRT* normalized to *TO* summarizing all cases of different chip sizes, scenarios of application mixes, and process variation maps for different aged states of the chips (i.e., in the initial unaged state and after 1, 5, and 10 years). A more detailed chip-level view of the box plot results is presented in Fig. 7.4. RPF of 100 denotes the ideal case with full protection and no deadline misses, which is hard to guarantee in resource-constrained scenarios because the number of cores is less than the required number to fulfil the RMT requirements of all the concurrently executing tasks. Moreover, in the presence of process variations, some cores will operate at a low frequency and thus it may not be feasible to ensure 0 % deadline misses for all the concurrently executing tasks. The evaluations for a wide set of values for different parameters (as discussed above) cover a broader range of use cases and resource-constrained scenarios.

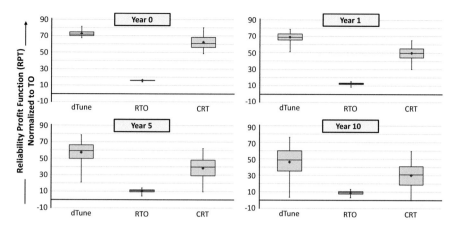

Fig. 7.3 Box plot results showing the overall RPF Improvements of dTune, RTO, and CRT Normalized to TO [Each *box plot* shows the summary of 31,500 experiments considering all cases of different chip sizes, scenarios of application mixes, and process variation maps over different aged states of the chips]

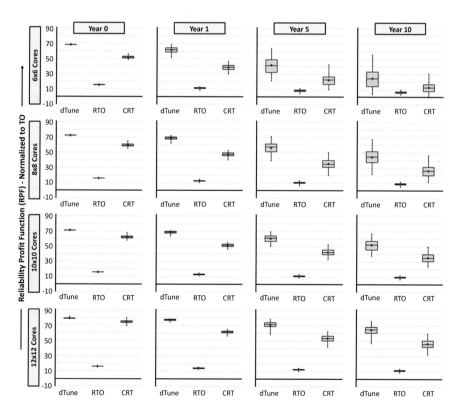

Fig. 7.4 Individual box plot results of RPF improvements of dTune, RTO, and CRT normalized to TO for different chip sizes and different aging years [Each *box plot* shows the summary of 3500–12,000 experiments considering all cases of application mixes and process variation maps]

In summary, for the year 0 (i.e., no aging) considering core-to-core frequency variations due to only process variations, the proposed cross-layer *dTune* system provides *RPF* savings of 57 and 11 % on average compared to *RTO* and *CRT*, respectively. For the aging year 10, compared to *RTO* and *CRT*, the dTune system achieves *RPF* savings of 37 and 16 % on average, respectively. In the following, detailed discussions are provided for individual comparisons.

Discussing dTune vs. TO at Year 0: In general, the reduced *RPF* in case of *TO* is given by the fact that the *dTune* system considers reliability and timing in a combined way instead of focusing on timing only. The minimum and maximum *RPF* improvements of *dTune* over the *TO* technique are due to the dynamic resilience-aware RMT tuning decisions, i.e., the highly vulnerable tasks get the cores first for redundant thread execution which is in particular beneficial in cases of application mixes with a large number of concurrently executing tasks. In case the system load is high, the RMT for highly resilient application tasks (i.e., task with low FVI values) is deactivated in order to facilitate other tasks and to preserve the timing reliability. As a result, *dTune* shows a significant improvement compared to *TO*. Note, a full protection without deadline misses cannot be achieved in many cases, as for a higher number of application mixes being executed the number of cores available is not sufficient to execute all tasks in the redundant multithreading mode. The minimum *RPF* values correspond to the cases of chips that have a higher number of cores with low operating frequency leading to more deadline misses and a reduced potential for reliable code version upgrades. The spread of the box plot is concentrated as most of the chips offer a possibility for RMT activation for the highly vulnerable tasks and version upgrades without deadline violations. Compared to *TO*, *dTune* achieves an average reliability improvement of 73 % (maximum 82 %).

Discussing dTune vs. TO at Year 1, 5, and 10: In the later years when cores are aged, the minimum reliability improvement of *dTune* compared to *TO* decreases significantly compared to year 0. It is attributed to the aging of the slower cores of a subset of chips resulting in the decrease of the processing capabilities of the chip that lead to more deadline violations, i.e., degradation of the timing reliability. Note that the aging degradation from year 0 to year 1 is higher than, e.g., between year 1 and 5. However, the maximum and average improvements are only slightly reduced that illustrate the benefits of considering knowledge from multiple system layers and adaptive RMT tuning for tasks with variable resilience properties. The spread of the box plot between maximum and minimum increases noticeably when core aging proceeds. It can be attributed to the lower optimization potential for the chips with an increased number of cores with low operating frequencies that are further degraded due to aging.

Discussing RTO vs. TO: The overall improvements of *RTO* compared to *TO* show the potential of the joint consideration of reliability and timing. However, as the *RTO* technique only uses the compiler layer and does not account for the knowledge from the hardware and redundant multithreading, its *RPF* improvements are significantly smaller than that of the *dTune* system. The spread of the box plot for *RTO* is very low as for most of the chips the reliability maximizing function versions can be selected without violating the deadline. Only for a limited set of chips either relatively more vulnerable versions for several tasks have to be selected in order to avoid deadline misses *or* the deadline is violated when the decrease in vulnerability offered by a version provides a better tradeoff between reliability and timing. For the aged chips (i.e., at year 1, 5, and 10), the average *RPF* improvement decreases while the spread between minimum and maximum increases due to the reduced potential of version upgrades *or* increasing deadline misses that are caused by the changed chip characteristics. It can also be noticed that the difference between year 5 and 10 is low due to the flattening of the aging curve. In general, however, *RTO* performs better than *TO*, but it is significantly worse compared to the *dTune* system due to the ignorance of the hardware-level knowledge.

Discussing dTune vs. RTO and CRT: *RTO* benefits from the reliable code version adaptation but the potential decreases due to ongoing aging. Compared to *RTO*, the *dTune* system still benefits from variation- and aging-aware core assignment and resilience-driven dynamic activation/deactivation of RMT, i.e., exploiting the low-level hardware knowledge. As a result, *dTune* achieves significant reliability improvements compared to the *RTO* technique, i.e., on average 57 % *RPF* improvement. For year 10 the minimum of *dTune* and *RTO* are similar, however the maximum and average of *dTune* is significantly higher showing that *dTune* is capable of adapting to the different characteristics of the chips.

Compared to *CRT*, for only few applications, *CRT* and the *dTune* system show a similar behavior. Highest cases of reliability savings denote the execution scenarios where sufficient cores are available to facilitate RMT for all the concurrently execution applications in the *TMR* mode with the fastest version. However, *CRT* does not adapt to an increasing number of tasks where timing and reliability-based selection need to be taken into account. Due to availability of multiple reliable code versions from the compiler layer, the *dTune* system has an additional benefit for matching core variations with soft-error resilient code versions. The minimum RPF improvement for the *CRT* technique in year 10 is lower compared to that for the *dTune* system and the *RTO* techniques so that *CRT* does not provide an improvement over *TO* any more. It is due to the fact that the aged cores are not accounted for because still all tasks are executed with RMT which leads to a significant number of deadline misses. This lower ability of *CRT* to adapt to the different chip characteristics (regarding the avoidance of deadline misses) is also reflected in the relatively high spread between minimum and maximum improvements in Year 0. This can also be seen when analyzing the maximum of *CRT* for year 1, 5, and 10. While for the unaged state of the chip the maximum improvement is almost as high as the one achieved by *dTune*, for the following years the number of deadline violations increases significantly for *CRT*. The decrease in the average improvements for *dTune* is lower than that of *CRT* when comparing year 0 and year 1. Even though *dTune* does not activate RMT for all tasks when the system load is high, the better average improvement compared to the improvement of *CRT* shows that the combination of the RMT selection for most vulnerable tasks and the possibility to select less vulnerable versions for tasks not being executed in RMT is effective, and thus demonstrating the benefits of a cross-layer reliability optimization technique. Overall, compared to the *CRT*, *dTune* achieves an average reliability improvement of 37 %.

Comparing *CRT* and *RTO* the improvements of *CRT* are significantly higher, especially for the unaged chips. The reason for this behavior is that *RTO* does not consider running tasks in RMT which would be possible for a low number of application mixes without a significant increase in the number of deadline violations.

Figure 7.4 shows individual box plots for different chip sizes and different aging years, where each box plot shows the *RPF* improvements of *dTune*, *RTO*, and *CRT* normalized to *TO* summarizing all cases of application mixes and process variation maps. Note that the general observations described above for Fig. 7.3 hold true for Fig. 7.4 as well. However, additional observations can be made when analyzing the box plots for different chip sizes individually. In general, in architectures where more cores are available the minimum, maximum, and average reliability improvements compared to the *TO* technique are higher for all approaches. The reason for that is that even if the application mixes are growing in proportion to the number of cores available, for the lower and medium range of application mixes, there are more candidate cores with higher performance, that can be taken advantage of in order to avoid deadline misses (in *RTO*, *dTune*), for RMT activation (in *dTune*, *CRT*), and for version selection (in *RTO*, *dTune*). In summary, for the year 0 (i.e., no aging), the proposed cross-layer *dTune* system provides *RPF* savings of 5–63 % for different chip sizes compared to different comparison partners. For the aging year 10, *dTune* provides *RPF* savings of 12–18 % for different chip sizes compared to different comparison partners.

7.5 Detailed Comparison Results

Figures 7.5, 7.6, 7.7, and 7.8 show detailed bar graph plots for different chip sizes (i.e., 6×6, 8×8, 10×10, and 12×12) with various process variation maps considering different application mixes with varying number of parallel executing tasks and different aging years. Each bar shows the reliability improvements of *dTune*, *RTO*, and *CRT* normalized to *TO* averaged over all process variation maps. Again, the general observations, reasoning, and discussion points described above for Figs. 7.3 and 7.4 hold true for Figs. 7.5, 7.6, 7.7, and 7.8 as well and only additional observations are discussed in the following when analyzing the bar graph plots for different application mixes for different chip sizes and aging years individually. It is important to note that, for a given chip size, as the number of tasks grows the *RPF* savings decrease. This can be attributed to two facts: (1) not all tasks can be fully supported in the RMT mode and (2) insufficient fast cores are available to meet the deadlines of all the tasks. Adaptive RMT tuning, reliable code version selection and adaptations enable the *dTune* system to adapt to different scenarios of application mixes. The resilience-driven core allocation for redundant multithreading enables dTune to achieve higher reliability savings in the competitive scenarios. Furthermore,

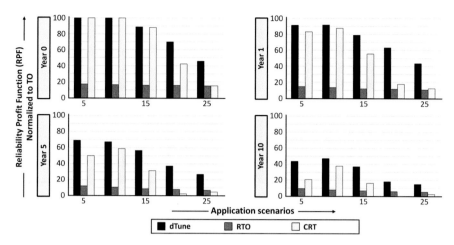

Fig. 7.5 Detailed reliability savings compared to different state-of-the-art techniques for the 6×6 core processors with different process variation maps considering different application mixes and aging years [Each *bar* corresponds to an averaged RPF saving value of 500 experiments considering different process variation maps]

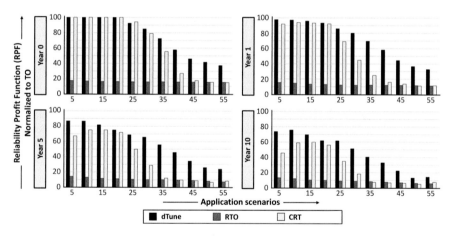

Fig. 7.6 Detailed reliability savings compared to different state-of-the-art techniques for the 8×8 core processors with different process variation maps considering different application mixes and aging years [Each *bar* corresponds to an averaged RPF saving value of 500 experiments considering different process variation maps]

variation-aware thread-to-core mapping facilitates meeting deadlines of timing critical tasks while accounting for the tasks requiring cores for RMT. In contrast to *dTune*, the *CRT* technique performs significantly worse as the number of concurrently executing applications increase. This is primarily attributed to unavailability of cores to fulfill the RMT requirements of all applications. The *RTO* technique stays consistent and low in terms of its reliability improvements due to the unavailability of the RMT feature and the only potential is using the

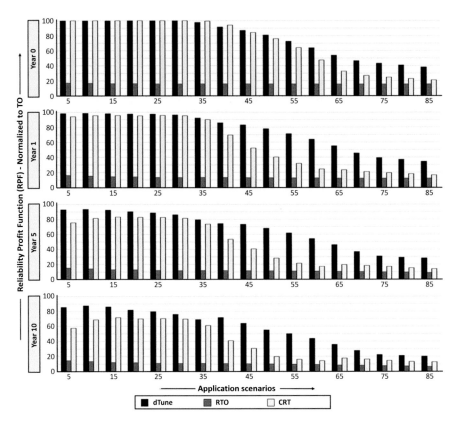

Fig. 7.7 Detailed reliability savings compared to different state-of-the-art techniques for the 10×10 core processors with different process variation maps considering different application mixes and aging years [Each *bar* corresponds to an averaged RPF saving value of 500 experiments considering different process variation maps]

reliable code versions trading reliability with timing. The trend of decreasing reliability profit of the *dTune* and *CRT* techniques with increasing number of applications stays consistent through different chip sizes and different years, while the *dTune* system being better than *CRT* in almost all cases. Compared to *CRT*, for a 6×6 chip, the reliability savings range from 0 to 31 %, 4 to 43 %, 8 to 34 %, and 9 to 21 % for year 0, 1, 5, and 10, respectively. The ranges of reliability savings for a 12×12 chip are -7 to 14 %, 2–34 %, 3–31 %, and 5–28 % for year 0, 1, 5, and 10, respectively.

Figures 7.5, 7.6, 7.7, and 7.8 illustrate that the reliability profit of the proposed *dTune* system improves or stays consistent across different chip sizes, multiple simulated years of chip aging, and a wide range of application mixes. In contrast, state-of-the-art techniques degrade severely due to the unawareness of applications' resilience properties, multiple reliable code versions, and low-level knowledge

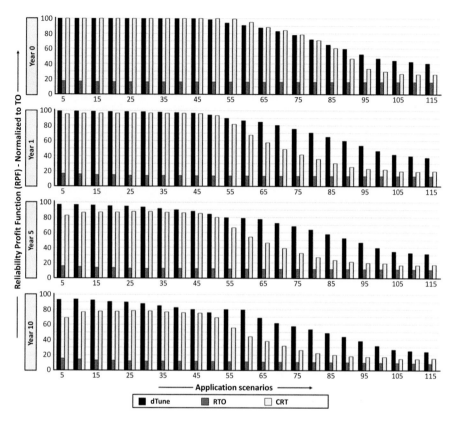

Fig. 7.8 Detailed reliability savings compared to different state-of-the-art techniques for the 12×12 core processors with different process variation maps considering different application mixes and aging years [Each *bar* corresponds to an averaged RPF saving value of 500 experiments considering different process variation maps]

about the aging-induced frequency degradation and variability during their core allocation decisions.

7.6   Detailed Analysis of Comparison Results for Two Chips for a 6×6 Processor

In the following section, a detailed analysis of the comparison results for two chips with different core-to-core variations (out of 100 chip process variation maps) is presented. The core-to-core frequency variations are shown for different years that are provided as an input to the proposed *dTune* system followed by the reliability savings. In the following, the results are discussed chip by chip.

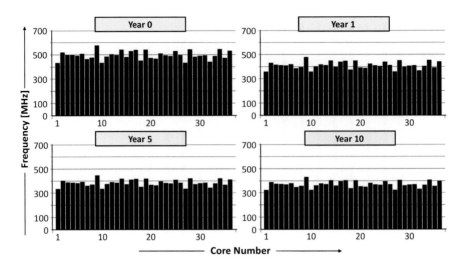

Fig. 7.9 CHIP19—size 6×6: core-to-core frequency distributions for different aging years

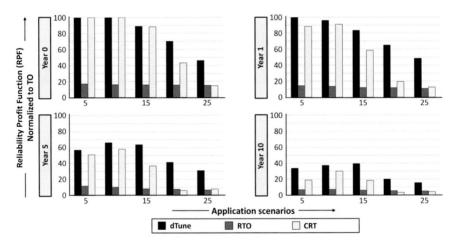

Fig. 7.10 CHIP19—size 6×6: detailed reliability savings compared to different state-of-the-art techniques for different application mixes and aging years

Chip 1: Figure 7.9 shows the frequency distribution of different cores for the case of a 6×6 core chip where the average frequency of the unaged cores (i.e., 496 MHz) is already lower than the nominal frequency (i.e., 500 MHz). After 10 years the average core frequency is only 371 MHz. In Fig. 7.10 the *RPF* improvements over *TO* are shown for the three different comparison partners. Although the average core frequency of the unaged chip is already lower than the nominal frequency, the savings compared to *TO* for *dTune* and *CRT* are still high. However, after 1 year, in case of *CRT*, the deadline violations increase, such that even for the five application mixes case, deadlines are violated, while *dTune* achieves full protection without any deadline violation. For 5 and 10 years, *dTune* and *CRT* both cannot avoid deadline

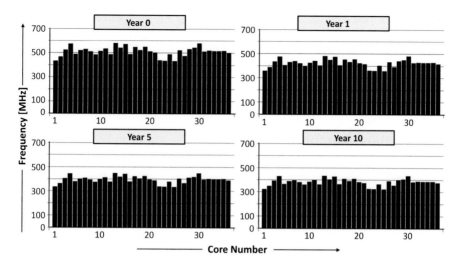

Fig. 7.11 CHIP41—size 6×6: core-to-core frequency distributions for different aging years

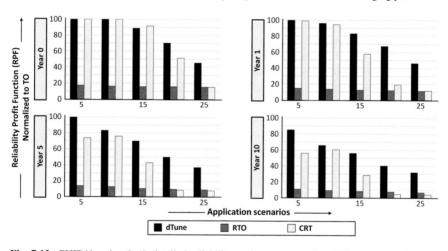

Fig. 7.12 CHIP41—size 6×6: detailed reliability savings compared to different state-of-the-art techniques for different application mixes and aging years

violations when activating RMT or executing highly reliable versions due to the decreased core performances. However, still a significant improvement can be achieved when comparing with *TO* for both years. For *RTO* it is also shown that with the ongoing aging, relatively less reliable versions can be selected, and the resulting improvement decreases.

Chip 2: Figure 7.11 shows the frequency distribution of the different cores for a different 6×6 chip where the average frequency of the unaged cores (512 MHz) is higher than the nominal frequency (500 MHz). After 10 years the average core frequency is only 382 MHz. Analyzing Fig. 7.12, both *dTune* and *CRT* achieve in general significantly better improvements compared to the cases in Fig. 7.10, especially for year 5 and year 10 due to cores having a higher performance. For the five application mixes case in year 5, *dTune* can still provide a full protection without deadline misses. This is also reflected in the results of *RTO*, where the improvements are higher compared to Fig. 7.10, as the faster cores allow reliability-wise better versions to be selected.

7.7   Chapter Summary

Compared to the single-layer approaches, i.e., Reliability-Timing Optimizing technique and Chip-Level Redundant Threading technique, the cross-layer approach provides savings of 57 % and 11 % on average for year 0 and 37 % and 16 % on average for year 10, respectively. When only considering the application-level knowledge of the different compiled versions in case of Reliability-Timing Optimizing technique, an improvement of 16 % compared to Timing Optimizing technique can be achieved. However, to achieve full soft-error resilience, redundant multitasking is important to be considered in combination with process variation and timing aspects. When considering only architecture-level knowledge (like in the Chip-Level Redundant Threading technique), in case not all tasks can be protected using redundant multithreading, the application-level knowledge of resilience is important and the process variation needs to be considered for synchronizing the outputs of redundant threads and to perform the voting without significant timing penalties. In such scenarios, the multiple reliable versions can also be exploited. Towards this end, the proposed Dependability Tuning system employs the information from different layers as mentioned in Sect. 7.3.1 in a cross-layer fashion. As a result it provides improved reliability efficiency (16 %–57 %) compared to different state-of-the-art single-layer techniques.

In summary, this work leverages multiple system layers and exploits the interaction between these layers to achieve an overall improved system reliability considering both functional and timing correctness under constrained scenarios, while state-of-the-art does not.

Chapter 8
Summary and Conclusions

Scaling down the transistor dimensions has led modern systems to become more and more susceptible towards various types of reliability threats such as soft errors, design-time process variation, and run-time aging effects. In an on-chip manycore system, different cores may become subjected to diverse reliability threats, thereby ignoring one threat while mitigating the other threat might lead to designing a system with reliability inefficiencies. This manuscript aims to mitigate the soft error issue at the software layers targeting unreliable (or partially reliable) embedded hardware while taking into account the run-time aging and design-time process variation effects. To enable this, a novel *cross-layer reliability analysis, modeling, and optimization* approach is proposed in this manuscript that leverages multiple layers in the system design abstraction (i.e., hardware, compiler, system software, and application program) to exploit the available reliability enhancing potential at each system layer and to exchange this information across multiple system layers. In order to achieve enhanced reliability in constrained scenarios, the proposed cross-layer reliability modeling and optimization flow employs various concepts and techniques at different software layers (i.e., where errors are observed) while accounting for the hardware-level knowledge (i.e., where faults occur) and software-level error masking properties.

The design of the cross-layer reliability modeling and optimization flow is enabled by a comprehensive reliability analysis of software programs when faults are injected in the hardware. This analysis helps in understanding how hardware-level faults manifest at the software layers and establishes a relationship between faults in different processor components and different types of instructions to different types of resulting errors. Furthermore, this analysis also exposes the potential of software-level masking, i.e., how different errors get masked due to data and control flow properties and do not get visible at the output of the application software program. Towards this end, one of the key novel contributions of this manuscript is to bridge the gap between hardware and software for cross-layer reliability modeling and optimization, i.e., considering the hardware-level knowledge while quantifying the effects of hardware-level faults at the software level. The accuracy of

© Springer International Publishing Switzerland 2016
S. Rehman et al., *Reliable Software for Unreliable Hardware*,
DOI 10.1007/978-3-319-25772-3_8

software-level reliability analysis is evaluated against the state-of-the-art SymPLFIED [98] approach where the latter provides 27 % overestimation of Application Failures w.r.t. *wrong access to Instruction Memory* due to an increased number of faults in the program counter. The program-level reliability analysis helped in identifying different parameters important from the software perspective, i.e., the notion of critical and noncritical instructions, spatial and temporal vulnerabilities, and the consideration of area and fault probabilities of different processor components for software reliability estimation. This analysis is leveraged to develop cross-layer software reliability models at various levels of granularity (i.e., instruction, basic block, and function/task) for quantifying the effects of hardware-level faults at the software level, while accounting for both the hardware- and the software-level parameters as discussed above.

The second key contribution of this manuscript is software reliability models to quantify three key important reliability aspects at the instruction granularity, i.e., (1) Instruction Vulnerability Index (IVI) that estimates the instruction's vulnerability to soft errors by jointly accounting for the spatial and temporal effects, (2) Instruction Error Masking Index (IMI) that estimates the masking probabilities for each instruction, and (3) Instruction Error Propagation Index (EPI) that estimates the error propagation effects at the software program level. These instruction-level reliability estimates are then used to obtain reliability estimates at the function/task levels. Furthermore, in order to account for the timing aspects, a Reliability-Timing Penalty (RTP) metric is introduced. A comprehensive analysis of IVI, IMI, and EPI is performed for various applications to demonstrate their varying reliability properties and relationship to their instruction profiles. In the cross-layer optimization flow, these reliability models are leveraged by different techniques at different system levels to quantify the reliability-wise importance of different instructions and functions/tasks and to perform selective/adaptive reliability optimization under tolerable performance overhead constraints.

This manuscript presents different concepts and techniques for *cross-layer reliability optimization* that leverage multiple system layers for reliable code generation and execution of application software programs. The aim is to first reduce the error probability by reducing the spatial and temporal vulnerabilities of instructions in different pipeline components and the number of critical instruction executions (for instance, load/store, call, branch, critical ALU instructions) through different reliability-driven software transformations. Under the scope of this manuscript, the following reliability-driven software transformations are proposed: (1) reliability-driven loop unrolling, (2) reliability-driven common expression elimination and operation merging, (3) reliability-driven data type optimization, and (4) reliability-driven online table value computation. Afterwards, a reliability-driven instruction scheduler determines the instruction execution sequence that influences the vulnerabilities of different instructions in different processor components. These techniques primarily aim at reducing the spatial/temporal vulnerabilities of instructions and number of critical instruction executions in order to reduce the probabilities of Application Failures and Incorrect Outputs. Applying these reliability-driven transformations and the instruction scheduler provides on average 60 % lower Application

Failures, hence enhancing the software reliability. The reliability is further improved by applying redundancy to only the critical instructions. Towards this end, a selective instruction redundancy approach is proposed which selects a set of reliability-wise important instructions (identified using the reliability models discussed above) in different functions for redundancy-based protection under user-provided tolerable performance overhead constraints. This enables a constrained reliability optimization where applying an expensive full scale redundancy to the software code parts is avoided. The key is to give more protection to the less-resilient part of the software program and less protection to the more-resilient part in order to achieve a high degree of reliability in constrained scenarios. Compared to state-of-the-art protection schemes [71, 76, 134], the proposed selective instruction protection provides on average 30–60 % improved reliability at 50 % tolerable performance overhead.

To enable run-time tradeoffs between the reliability and performance properties of applications, multiple reliable versions of a software program are generated using the proposed transformations and selective protection techniques. Once multiple reliable code versions are generated, reliable execution of the code is facilitated through a reliability-driven run-time system that takes into account the core-to-core frequency variations due to design-time process variation and run-time aging induced frequency degradation. Soft error resiliency is enhanced by adaptively activating/deactivating the Redundant Multithreading (RMT) for different applications while accounting for their distinct resilience properties and deadline requirements along with a history of the encountered errors under an area-constrained scenario. Depending upon the cores' frequency variations, a suitable reliable version for each application is selected. Finally, the task to core mapping is performed for redundant thread execution at run-time such that the execution properties of the redundant threads are closely matched to the frequency properties of allocated cores considering core-to-core frequency variations. Compared to state-of-the-art single-layer reliability optimizing techniques (i.e., CRT—Chip-Level Redundant Multithreading [36] and RTO—Reliability-Timing Optimizing Technique [141, 142]), the proposed cross-layer approach achieves 16–57 % improved software reliability on average for different chip configurations, various process variation maps, different aging years, and wide range of application mixes with concurrently executing applications.

In addition to the above-discussed scientific contribution, several tools for soft error analysis, aging analysis, and an integrated fault generation and injection system for instruction set simulators have been developed in the scope of this work and are made available at http://ces.itec.kit.edu/846.php. Besides the tools, this work has provided an initial foundation for a spin-off project under the name of "GetSURE" which is a joint collaborative project under the DFG-funded Special Priority Program (SPP) 1500.

In summary, a robust and dependable embedded system design needs to consider the reliability at all abstraction levels. This work develops a cross-layer reliability modeling and optimization flow that incorporates novel reliability-driven compiler, reliability-driven system software, and application-guided fault tolerant mechanisms to achieve effective reliability optimization under constrained scenarios. These techniques are enabled by accurate software program reliability models.

State-of-the-art software reliability schemes by principle cannot achieve the level of reliability that the proposed cross-layer approach can, because of the following conceptual differences:

- This work jointly accounts for the temporal *and* spatial vulnerabilities in the underlying reliability modeling and fault injection flow along with program-level error masking and propagation properties at different abstraction levels, i.e., instruction and function level (state-of-the-art techniques do not).
- This work leverages multiple system layers and exploits the interaction between these layers to achieve an overall improved system reliability considering both functional and timing correctness under constrained scenarios, while state-of-the-art does not.
- The proposed techniques consider Application Failures based on soft errors observed when the software program is executed in a processor pipeline stage by stage (state-of-the-art techniques do not provide these details and therefore cannot optimize for it).
- State-of-the-art techniques try to incur no performance overhead or provide full redundancy with a very high overhead, whereas the techniques presented in this manuscript leave it up to the user to specify a tolerable performance overhead that may be associated with a gain in reliability. Often, a small performance loss may enable a high reliability gain. The proposed techniques can exploit those scenarios, state-of-the-art techniques do not.

Appendix A
Simulation Infrastructure

This chapter presents the integrated tool flow, simulation environment, and fault injection setup that are developed in this work. The reliability analysis and evaluations of the individual techniques compared to state-of-the-art approaches are already discussed in the previous chapters, i.e., Chaps. 4, 5, and 6. The complete overview of the developed tool chain and infrastructure is shown in Fig. A.1.

Based on a processor description (e.g., of LEON3) in the form of VHDL files and a technology library containing different gates, the processor is synthesized using the *Synopsys Design Compiler* in order to obtain the netlist and a set of critical paths. This data is then used to estimate the area and fault probabilities of different processor components, and the aging of the critical paths. Additionally, logic simulations using *ModelSim* are performed executing different applications on the synthesized processor for extracting their respective activity and signal probabilities. The information about the area and fault probabilities of different processor components, which is obtained after the processor synthesis, is used to estimate the program reliability. In the *reliability-aware manycore simulator*, the run-time aging estimation results are finally used jointly with the design time process variations (that are input into the infrastructure as variation maps) to account for varying performance characteristics of different cores. The simulation environment is based on an Instruction Set Architecture (ISA)-simulator for LEON3 cores generated using the ArchC tool chain [119]. For reliability analysis and estimation, different applications are simulated and the required data for devising the models and for parameter estimation is obtained. Furthermore, the simulator is equipped with a configurable fault generator, fault injector, and error logging modules for characterizing the impacts of the reliability threats on different applications. Different benchmark applications from the MiBench benchmark suite [111] are used for evaluation, which form the input to the reliability-driven compilation setup that generates versions with different reliability and performance characteristics by employing different techniques. In the following, the individual parts and tools of the infrastructure are discussed.

© Springer International Publishing Switzerland 2016   193
S. Rehman et al., *Reliable Software for Unreliable Hardware*,
DOI 10.1007/978-3-319-25772-3

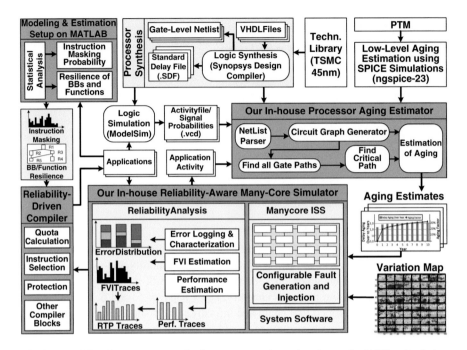

Fig. A.1 Tool flow for processor synthesis, processor aging estimation, and reliability simulation and evaluation for manycore processors

Several tools developed in the scope of this manuscript are made available online for download at http://ces.itec.kit.edu/846.php.

A.1: Reliability-Aware Manycore Instruction Set Simulator and Fault Injection

The reliability-aware Instruction Set Simulator (ISS) is based on the ArchC architecture description language and related tools [119] (described in Sect. A.2). It simulates a *SPARC v8* pipelined architecture with 16 kB of ECC-protected instruction and data caches; see area details in Table 7.1 and processor layout in Fig. 3.3. The simulation environment is extended with a configurable fault generation and injection module that injects faults in different processor components during the application execution; see different input parameters in Table A.1.

The fault rate (in #faults/MCycles) is obtained using the neutron flux calculator [126] and city coordinates which determine the geographical location and altitude where the device will be used. Considering various locations, we obtained three different fault rates in our experiments (1, 5, 10 faults/MCycles) to cover a wide range of cases (terrestrial to aerial), which conforms to the test conditions opted by

Table A.1 Different parameter for fault scenario generation

Parameter	Description	Properties/values
Distribution	Distribution models for fault generation	Random, uniform
Bit flips	Min/Max number of bits flipped	1/1, 1/2, 1/3, …
Fault probability	Probability that strike becomes a fault	Output of Sect. 4.1.3
Fault location	List of target processor components	Register file, PC, IW, IM, DM, etc.
Processor layout/area	Size of the complete target device	in mm² (Output of Sect. 7.1.1)
Component area	Area of different processor components given as percentage of processor area	0–100 %
Place and altitude	City and altitude at which the device is used to determine the flux rate	Karlsruhe, Germany; 1–20 km
Frequency	Operating frequency of the processor	100–500 MHz

prominent related work [77, 80] and as such eases comparison. The errors are observed at the application software layer and are classified in different error categories. Numerous fault injection campaigns were performed with different configurations like flux rate, operating frequency, fault models (single or multiple bit flips), and distribution models; see fault injection parameters in Table A.1. The complete methodology of the reliability-aware simulation and analysis is done in two major steps: (1) fault generation and injection during simulation and (2) error analysis and estimation.

A.2: ArchC Architecture Description Language

ArchC [119] is an Architecture Description Language (ADL) which is based on SystemC and is used to define processor architectures following the C++/SystemC syntax style. ArchC facilitates the designers to model new architectures and also to experiment with existing ones. Furthermore, the architecture can be described on various abstraction levels (e.g., functional or cycle-accurate description of an architecture) and afterwards the generation of software tools (e.g., simulators, assemblers, and linkers) can be performed automatically. For various architectures, the ArchC descriptions are available in [127], for instance, MIPS, Intel 8051, and *SPARC v8* (which is adopted for evaluation in this research and described in the following), that can be used to generate functional simulators. Furthermore, ArchC offers a co-verification mechanism that enables checking the consistency of a refined model against a reference model.

For the simulator generation two basic input descriptions are required:

- *Architecture Resources (AC_ARCH)*: the information regarding the resources, e.g., pipeline structure and memory hierarchy needs to be defined (see *sparcv8. ac* for the *SPARC v8* architecture).

- *Instruction Set Architecture (AC_ISA)*: details about every instruction such as the format, opcode, and behavior need to be described (see *sparcv8_isa.ac* for the instruction declarations).

These descriptions serve as an input to the *ArchC Simulator Generator* (acsim), which outputs the C++ classes and SystemC modules required to build the simulator. Additionally, the ArchC Simulator Generator uses a decoder generator and a preprocessor for lexical analysis and parsing of the language, which extracts information from the description files. The following files are created; only the important ones for a *SPARC v8* architecture are listed below.

- *main.cpp*: this file provides the facility to instantiate the model and several features can be set here. It can be extended for the usage of additional SystemC modules.
- *sparcv8.cpp*: the processor module is implemented in this file. Amongst others, it contains a loop in which the decoding and the appropriate instruction behavior are called.
- *sparcv8_isa.cpp*: the behavior description for all instructions of the *SPARC v8* architecture is presented here. This file is created as a template and the behavior method for every instruction is placed inside by the designer. The description of an instruction behavior comprises of a general instruction behavior, which is common for all instructions, a format behavior, which is common for all instructions that have the same instruction format, and a specific instruction behavior for each individual instruction.

Figure A.2 shows the complete flow for the ArchC simulator generation. A GCC compiler is used to compile (i.e., by running "*make -f Makefile.archc*") the created model or to extend the existing ones and produces an executable specification of the target architecture. The generated simulator executes instruction decoding (that can be speeded up using a cache for decoded instructions), scheduling and behavior dynamically. Moreover, it supports operating system (OS) call emulation so that it is possible to simulate applications which contain I/O operations.

This step outputs a file *sparcv8.x* which is used for an application simulation that has been compiled using the automatically generated tools. Further detailed descriptions of ArchC and the related tools can be found in [119] and [128].

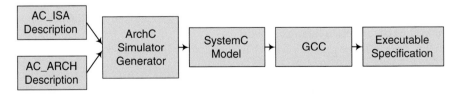

Fig. A.2 ArchC simulator generation [128]

A.3: Reliability-Aware Simulation and Analysis Methodology

Figure A.3 shows an overview of the developed simulation and analysis methodology for evaluating different software program reliability techniques. It works in two main phases that operate in an automated flow.

Fault Injection and Simulation Phase: The fault injection technique integrated in the instruction set simulator (ISS) is equipped with a configurable fault generation engine (described below in detail). The fault generator generates different fault scenarios considering different fault models (e.g., number of bit flips and distribution), fault rates, and faults in different architectural components (e.g., register file, Program Counter (PC), Instruction Word (IW)). Processor-specific details (chip footprint, component area, number of registers, etc.) and fault model configurations are passed as input (Table A.1).

The number of injected faults per component is determined by the component area (obtained after RTL synthesis) to incorporate spatial vulnerability. For example, fewer faults are injected in the PC compared to the ALU/Multiplier. The fault modeling procedure at the ISS level is illustrated in Fig. A.4. For example, a fault in the instruction decoder or in the IW is modeled as corrupting one/multiple fields of the IW in the ISS that results in a wrong opcode or wrong operand. The faults are injected during the application execution. If a fault is injected into the multiplier while an *add* instruction is being executed, it will have no effect on the application output. Note that the modeling procedure and fault injection are generic and independent of a particular architecture implementation. In case of a protected component, the correct state is resumed immediately after the fault injection. In the following, the fault generation and fault injection steps are explained in more detail.

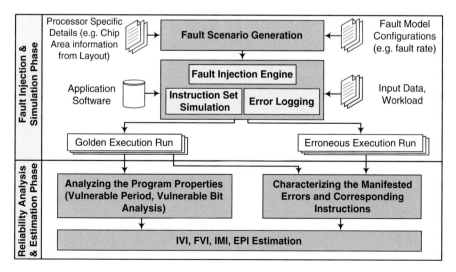

Fig. A.3 Flow of the reliability-aware simulation and analysis

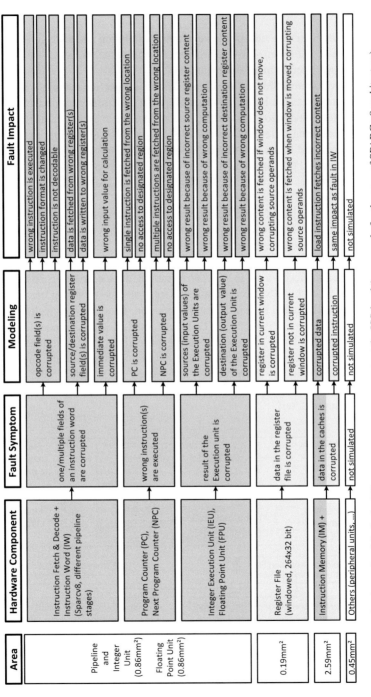

Fig. A.4 Modeling hardware-level faults in different processor components at the ISS-level (an example for the case of SPARC v8 architecture)

[Cycle]:	,[Duration],[Type],[Location],[Vector],[Address],;
Cycle:	In which cycle the fault should be injected
Duration:	How long the fault should stay
Type:	Type of fault (e.g. transient fault, stuck-at fault, etc.)
Location:	In which component the fault should be injected
Vector:	# of bits and their positions for bit flips
Address:	sub-address of fault location (e.g. register number in case of the register file)

226668:	,1,1,6,131074,2,;
4458402:	,1,1,4,32768,0,;
5271986:	,1,1,3,65602,227,;
94276206:	,1,1,1,71680,1,;
⋮	

Fig. A.5 Format and an excerpt of a fault file

Configurable Fault Generation Engine: This component generates a set of *fault files*, that contain the information about the faults to be injected later (see Fig. A.5 for an excerpt) providing details on *when* (i.e., in which cycle) and *where* (i.e., in which processor component) a fault is to be injected. The fault generation module works independent from the fault injection module. The reasons for this are: (1) to reuse the same fault scenarios for different applications for comparison and for reproducibility of the results and (2) to extend (if required) the fault generation module with additional parameters. Depending upon the input test conditions (e.g., number of bit flips and fault rates), the input parameters are configured.

Once the Fault Generation Engine finishes its execution, it outputs a *fault file* which comprises of two parts: (1) a header, which summarizes the information regarding the configuration settings for the fault file generation module and (2) the content, which shows the detailed information regarding the faults, where each line has the cycle information, i.e., the start cycle of a fault. The fault injection related entries are divided into blocks of five comma-separated fields. Each block represents one fault with its duration, type of fault, the fault location (i.e., the component in which the fault should be injected), a fault vector which specifies the number and the exact position of the bits where the faults are injected, and a specific address within a fault location (e.g., memory/register address) where the faults are injected. Figure A.5 shows the content section of the *fault file*. Numerous fault files are generated representing different scenarios and configuration settings using a script. For every script execution a separate directory is created which contains a set of fault files. These fault files are finally given as an input to the fault injection engine that injects faults during the program execution.

Fault Injection Engine: The step-by-step operational flow of the fault injection engine is shown in Fig. A.6. The compiled application versions are executed on an Instruction-Set Simulator which is enhanced with the capability to trace the application execution. During the simulation of the application, the fault scenarios are applied using a fault injector and the errors triggered are logged. For fairness of comparison and reproducibility, the same fault scenarios are used for the evaluation of all applications and their versions. The results of the fault injection experiments are obtained later by analyzing the effects of hardware level faults on the application software program level for each individual simulation. Afterwards, the program output errors are categorized and the reliability for the different application versions is computed using different reliability metrics. The following steps are taken:

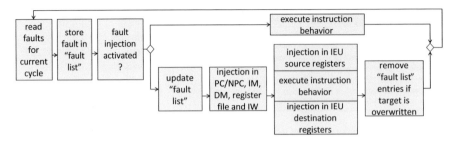

Fig. A.6 Flow of the fault simulation process

1. The faults for the current cycle are read from the fault file and are stored in a *fault list* which contains all faults that are currently injected.
2. If the fault injection is activated (a functionality implemented to be able to inject faults only in specific functions/parts of a program), the *fault list* is updated and bit flips are injected in the respective components.
3. Afterwards the current instruction is executed.
4. Then, the *fault list* is inspected for entries whose target is overwritten. Those entries are removed from the *fault list*.
5. If the fault injection is deactivated, only the instruction is executed.

Note that the fault injection does not introduce any unwanted side effects like changing performance counters. The application program is additionally simulated without fault injection to obtain a "golden run" (i.e., correct execution). It is later used for comparison with the "erroneous run" to identify the potential errors in the program output.

Error Analysis and Reliability Estimation: An error analysis is performed for application reliability analysis while considering the application properties (e.g., histograms of the executed instructions). The error characterization and the properties of an application are used to obtain reliability metrics at different levels of granularity (i.e., the instruction and function/task level), which are used to quantify the susceptibility of an application program towards Application Failures.

Different scripts are used to automatically analyze the results of the application simulations that output a set of files, i.e., application output, a log file containing a trace of all executed instructions and summary file containing the execution time of the application and potential warning/error messages. Afterwards, the set of files obtained from the fault injection simulation of an application are compared to the set of files obtained after the fault-free simulation, i.e., the "golden run" output and the results are grouped into different categories. The category Correct Output is assigned in case the application terminates successfully and the output matches with that of the "golden run." In case the faulty and fault-free application simulations produce different or no output, a more detailed error analysis is presented in Sect. 3.2. In case of an Application Failure, an abnormal termination has to be detected. Furthermore, an error or a warning message can be seen inside the summary file. However, the log file is required to be analyzed for some special subcategories of

Application Failures, e.g., in order to identify the reason for a Segmentation Fault, the last executed instruction is required to be identified that is responsible for this type of Application Failure. In case the application is not terminated in an abnormal way, the output files have to be inspected in detail. In case the comparison shows incorrect data in the output, the category Incorrect Output is labeled.

The same procedure is repeated to generate all the simulation results and their outcome, i.e., the error distribution categories, which are produced on the basis of the created fault files with the same configuration settings and are stored in a comma-separated list. This format makes it easier to use the data for plotting the results in a graphical representation. The error characterization distribution shows the impacts that are caused by the injected faults. However, the reason for a certain application behavior (i.e., erroneous/error-free category) cannot be explained with this; therefore, a thorough analysis of the application source code and the log file is required in order to explain the reasons for a certain error category. The log file produced after the golden run is inspected to obtain the application characteristics, i.e., instruction profile, which is computed by counting the number of times each instruction type is executed. For completeness, the instructions are categorized: *call/branch/jump*, *sethi/nop*, *load*, *store*, and *logic and arithmetic* instructions. Furthermore, an average instruction profile over all executions can also be generated for a certain function. This information can be used to decide if an application is more data dominant or more control-flow dominant, and also provides an insight about the usages of the hardware components (e.g., multiplier, ALU). Additionally the value of the susceptible time for each register is calculated, which is the sum of the times between a register write and the respective last read access.

Performance Evaluation of the Developed Reliability Analysis Methodology: The reliability analysis methodology and fault injection experiments were performed using a 24-core (2.4 GHz) Opteron processor 8431 with 64 GB memory. The average performance of our fault injection and simulation is 72×10^3 SIPS (simulated instructions per second) with extensive error logging (40 MB/MCycles). The performance of the SymPLFIED [98] program-level fault injection approach is 15.2 SIPS. It shows that the proposed reliability analysis methodology offers significant performance improvement (>4K times) compared to the state-of-the-art approach, i.e., SymPLFIED, which is primarily due to the extensive model computation in SymPLFIED.

Figure A.7 illustrates the comparison of the reliability estimation accuracy of the proposed methodology and tool flow with SymPLFIED [98]. This comparison illustrates the benefits of bridging the gap between the hardware and the software to obtain accurate reliability analysis. It can be observed that in case of SymPLFIED, the number of application software *crashes due to wrong access to Instruction Memory* increases significantly. This is because of an increased number of faults that were injected in the PC. The main reason is the ignorance of the processor layout with several architecture-specific features in SymPLFIED's machine model. Therefore, the percentage of fault in the PC increases from 0.1 to 7.1 %, which leads to an average 27 % overestimation of Application Failures. The comparison analysis in Fig. A.7 demonstrates that when using the SymPLFIED technique, the

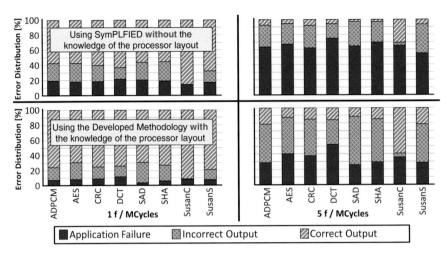

Fig. A.7 Detailed error characterization in different applications using our methodology and SymPLFIED [98]

probabilities for Application Failures and Incorrect Outputs are overestimated, which lead to an inaccurate FVI estimation. *It thereby demonstrates the improved accuracy of the proposed reliability analysis methodology and tool flow.*

Reliability-Driven Compilation: In order to compile the applications and their different versions, the GNU Compiler Collection (GCC) [129] is used. In this work, GCC is used due to its compatibility with the ArchC tool chain. In GCC several optimization options exist ranging from "O0," which enables a fast compilation and expected debugging results, to "O3," where many optimizations are activated that target performance improvement but lead to a longer compilation time. Besides that, the optimization option "Os" targets a reduction of the code size. Consequently, it is possible to create different application versions using the basic optimization options of GCC, whose impact on the software reliability can be analyzed using the presented analysis models and tools. The compiler-level reliability optimizing techniques are implemented at the source-code or assembly level or using the CDFG. The reliability analysis and estimation are done on CDFG or assembly code. In the following, potential ways for automatic application inside the integrated compiler flow are discussed.

Besides the standard optimization options for the *Reliability-Driven Software Transformations*, several additional, already existing compiler optimization passes can be used to analyze their impacts on the reliability of an application. Different application and function versions can be generated by activating/deactivating certain (optimization) options of the compiler. In GCC this can be done by either using "–f{optionName}" for activation or "–fno-{optionName}" for deactivation of an option. For example, the loop unrolling can be turned off using "–fno-unroll-loops." Additionally, several compiler constants/parameters can be changed using "– – param name=value," e.g., by setting "max-unroll-times" to a certain value the maximum amount of unrolling of a single loop can be defined [129]. Consequently, taking the example of the Reliability-Driven Loop Unrolling, the different versions that are used for the fault injection analysis can either be (1) generated automatically using GCC (using the above mentioned options and parameters) or can be (2) implemented in a high-level language at the source-code level.

To enable the *Selective Instruction Protection* and *Reliability-Driven Instruction Scheduling*, two alternatives can be selected: (1) After the compilation stage is finished or in case the high-level language source code is not available, the assembly code can be used as an input for, e.g., duplicating certain instructions, changing register allocations, and adding check instructions based on the instruction vulnerabilities. Afterwards, the modified assembly code can be assembled and linked. (2) The compiler can be enhanced by adding an additional optimization pass with additional input data, e.g., the instruction vulnerabilities. As an alternative, the vulnerability model at the required granularity can also be integrated in the compiler, e.g., using information on instruction dependencies and register allocation, as a static vulnerability estimation at compile time. Taking the example of the Reliability-Driven Instruction Scheduling, the basic block separations and the corresponding branch probabilities available in GCC can be taken advantage of.

Appendix B
Function-Level Resilience Modeling

In this appendix, a function-level resilience modeling technique is presented that was developed in the scope of this manuscript. The proposed resilience model quantifies the resilience of a given application function against the hardware-induced errors. This model can be used for characterizing the reliability importance of different functions and employing function-level reliability optimization techniques.

B.1: Definition

The *resilience* of an application function is defined as the probabilistic measure of functional correctness (output quality) in the presence of faults.

B.2: Modeling Function Resilience

Modeling resilience requires error probabilities for basic blocks outputs. There are two possible error types: Incorrect Output and Application Failure. Therefore, output of each instruction in a given basic block can be modeled as a Markov Chain with three states: S_C, S_{IC}, and S_F denoting Correct Output, Incorrect Output, and Application Failure states, respectively (see Fig. B.1). Considering that the execution of a program is a stochastic process, we adopt the Markov Chain technique for output modeling as it provides a fair tradeoff between the model complexity and accuracy when compared to exhaustive Monte-Carlo Simulations, Fault-Tree Analysis, and Principal Component Analysis based reliability models.

Assuming that each state is dependent upon the previous instructions' output and the error state can only be observed at the end or at the time of Application Failure, the execution path can be modeled as a Hidden Markov Chain, with the above-discussed

© Springer International Publishing Switzerland 2016
S. Rehman et al., *Reliable Software for Unreliable Hardware*,
DOI 10.1007/978-3-319-25772-3

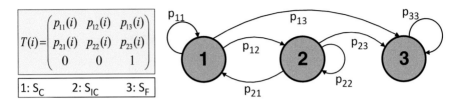

Fig. B.1 Markov Chain for instruction output with state transition probabilities

three states as hidden states and the observation state as "application failed" or "not-failed." The parameters of this model are the state transition probabilities as given in the matrix T and shown in Fig. B.1. These probabilities depend upon the executed instructions. Note, the Markov Chain is non-homogeneous as the transition probabilities change depending upon an instruction I_{ijk}.

After these probabilities are estimated (see parameter estimation later in this section), we can then compute the final state probability for a given basic block B_{ij} using Eq. A.1, where ξ is the final state probability vector containing the probability of three states: p_C, p_{IC}, and p_F.

$$\xi\left(B_{ij}\right)=\left[\begin{array}{ccc} p_C & p_{IC} & p_F \end{array}\right]_{B_{ij}} = \xi\left(B_{ij-1}\right)\times \Pi_{x\in I_{ij}}T\left(x\right) \qquad (A.1)$$

Following the information theory concepts, *the resilience of a function is modeled as the normalized mutual information between the required correct result (from a golden execution run X) and the result at the end of a function execution* (from a potentially faulty execution under a given fault rate), i.e., amount of useful function output. Mutual information is a measure of the amount of correct/useful information that can be inferred from the true result of the function/basic block (Fig. B.2 explains this concept). A large value of mutual information illustrates that more information about the correct output can be inferred, i.e., high resilience.

The mutual information between an always correct execution X and a real execution Y that may have some errors is represented as $I(X;Y)=H(X)-H(X|Y)$. The function $H(X)$ is the information obtained from the correct execution, which is 1 since the correct execution contains all the information possible. The conditional entropy is the information lost $H(X|Y)$ out of $H(X)$ which is the correct information. These concepts are used to quantify the resilience of a basic block B_{ij} which can be computed as $R(B_{ij})=1-H(X|Y)/H(X)$, where $H(X)$ is the information about the correct execution, i.e., $H(X)=b_{Live}$, where b_{Live} denotes the bits of live output registers of B_{ij}.

The conditional entropy $H(X|Y)$ is now the information lost in B_{ij} and given as Eq. A.2, where $p_C(x)$ represents the probability of correct value being x; and $p_{[IC,F]}(x, y)$ is the conditional probability of faulty output being Incorrect Output or Application Failure.

Fig. B.2 Flow for estimating the mutual information for function resilience

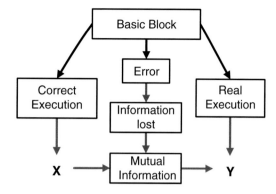

$$R\left(B_{ij}\right)=1-H\left(X\,|\,Y\right)/\,H\left(X\right); \quad H\left(X\right)=b_{\text{Live}}$$

$$H\left(X\,|\,Y\right)= \sum_{x\in X, y\in Y} p_{[\text{IC,F}]}\left(x,y\right)\times\log_2\left(p_C\left(x\right)/\,p_{[\text{IC,F}]}\left(x,y\right)\right) \qquad \text{(A.2)}$$

Assuming, resilience of a basic block $R(B_{ij})$ can be characterized as resilience to Incorrect Output and resilience to Application Failures, we can compute the conditional entropy separately for both cases. $H(X|Y)_F$ is given as $p_F(B_{ij})$ using Eq. A.1, while $H(X|Y)_{\text{IC}}$ is given by Eq. A.3.

$$H\left(X\,|\,Y\right)_{\text{IC}}=-\left[p_{\text{IC}}\times\log_2\left(p_{\text{IC}}\,/\left(2^n-1\right)\right)+\left(1-p_{\text{IC}}\right)\times\log_2\left(1-p_{\text{IC}}\right)\right]_{B_{ij}} \qquad \text{(A.3)}$$

By replacing the terms of Eq. A.2 with Eq. A.3, we can compute the resilience of a basic block against Application Failures and Incorrect Outputs where the second term in Eq. A.4 denotes the combined information loss.

$$R\left(B_{ij}\right)=1-\left[H\left(X\,|\,Y\right)_{\text{IC}}+H\left(X\,|\,Y\right)_F-\left(H\left(X\,|\,Y\right)_{\text{IC}}\times H\left(X\,|\,Y\right)_F\right)\right]/\,H\left(X\right) \qquad \text{(A.4)}$$

Given the resilience values of all basic blocks B_i of a function f_i, resilience $R(f_i)$ can be computed using Eq. A.5.

$$R\left(f_i\right)= \sum_{\forall b\in Bi}\left(R(b)/eF_i\left(b\right)\right)\times \sum_{\forall b\in Bi, \forall fi\in F}\left(eF_i\left(b\right)\right) \qquad \text{(A.5)}$$

Parameter Estimation: For estimating the model parameters, i.e., transition probabilities given in Eq. A.1, a few assumptions are made:

1. Observation of faulty output is made at the end of function
2. No recovery mechanism and no error protection is available, i.e., starting from a base case of unreliable hardware $\Longrightarrow p_{33}=1$; $p_{21}=0$.
3. Initial state and input is error-free; $[p_C\,p_{\text{IC}}\,p_F]_{(t=0)f1}=[1\ 0\ 0]$.

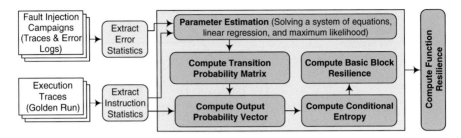

Fig. B.3 Flow of steps to compute basic block and function resilience

Moreover, $p_{11}+p_{12}+p_{13}=1$ and $p_{21}+p_{23}=1$. To expedite the parameter estimation process, the instructions are grouped into N_T primitive instruction categories (like arithmetic, multiply, divide, logical, load/store, calls/jumps, and floating point) such that all instructions in a given category share the same transition probabilities. The parameters can be estimated through extensive fault injection campaigns. Consider there are N_S different fault-injection experiments at a given fault rate, N_C and N_{IC} are the number of cases with Correct Output and Incorrect Output, respectively. For a particular fault injection experiment s, for a certain instruction category t_k, the transition probability p_{11} can be estimated using the maximum likelihood, thus deriving Eq. A.6. NI(t,s) denotes the number of instructions of type t in simulation s.

$$\log\big(p_{11}(t_k)\big)=-\text{NI}(t_k,s)\times\left(\frac{\displaystyle\sum_{\forall s\in S}\log\big(N_S/N_C(s)\big)+\sum_{t=0,t\neq ts}^{N_T}\text{NI}(t,s)\times\log\big(p_{11}(t)\big)}{\left(\displaystyle\sum_{\forall s\in S}\text{NI}(t_k,s)\right)^2}\right)$$

(A.6)

Assuming $p_{23}(t)=p_{13}(t)$, Eq. A.6 is utilized to obtain the probability $p_{22}(t_k)$. In this way all the remaining transition probabilities are computed, such that, $p_{23}(t_k)=p_{13}(t_k)=1-p_{22}(t_k)$; and $p_{12}(t_k)=p_{22}(t_k)-p_{11}(t_k)$.

Figure B.3 shows a simplified flow of different steps of our scheme towards modeling and estimation of function resilience along with parameter estimation and computation of conditional entropy.

Complexity: The complexity of resilience estimation is $O(|B_i|\times N_T\times\log(|I_{ij}|))$, which is much smaller than the complexity of fault tree based methods (i.e., $O(|B_i|\times|I_{ij}|^3)$) and Monte-Carlo simulations (i.e., $O(|B_i|\times|I_{ij}|^2)$) for each basic block.

B.3: Results

The resilience model quantifies the reliability properties at a coarse-grained level, i.e., function and basic block that can be used to facilitate in prioritizing different functions and basic blocks for selective protection/constrained reliability optimization. For example, allocating the performance quotas to different functions/basic blocks depending upon their higher/lower resilience values.

In this work the resilience is used as a metric to quantify the coarse grained reliability, i.e., at function/basic block level and then using the resilience values for distributing the tolerable performance overhead quota among different functions and different basic blocks inside the application program. A more resilient function would get a less quota for protection compared to a less-resilient function that may not tolerate more errors. Figure B.4 shows the resilience (in log scale) and the performance overhead quota for different application functions. The resilience and quota are provided separately for the Incorrect Output and Application Failure cases along with the combined case. Note, here Incorrect Output and Application Failure are both treated as information loss. Due to the high resilience, *DCT* gets lesser quota in comparison to the *ADPCM*, *SHA*, and *SAD*. The resilience of *DCT* is high because it is an unrolled version, with a relatively lesser number of branches, i.e., critical instructions, compared to other applications that lead to fewer control flow errors in *DCT*.

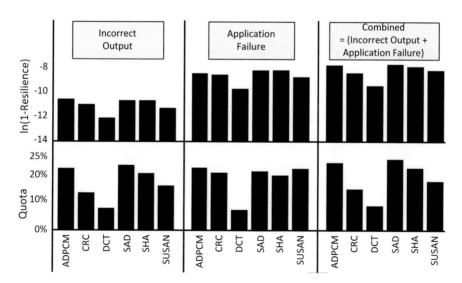

Fig. B.4 Resilience of various application functions (inverse values in log scale): resilience is shown separately for Incorrect Output and Application Failure, and *Combined*

Appendix C
Algorithms

C.1: Algorithm for Computing the Error Masking Probability PDP(*I*, *p*)

```
Error Masking Probability Computation
_____
Input: G (V, E), L_G, (P, S)
Output: Masking probabilities due to data flow for each instruc-
tion I for path p, P_DP (I, p)
1.     FOR all I ∈ G DO
2.         P_D (I) ← computeP_D (I); // Eq. 4.7
3.     END FOR
4.     FOR all I ∈ L_G DO
5.         P_DP (I, p) ← P_D (I) //  for all leaf nodes
6.     END FOR
7.     List L();
8.     FOR all x ∈ L_G.P DO
9.            L.add(x); //  list of ready nodes
10.    END FOR
11.    WHILE (!L.isEmpty()) DO
12.            FOR all I ∈ L DO
13.                I.Paths ← generatePaths(I);   //  generate
                   all instruction paths
14.                FOR all p ∈ I.Paths DO
15.                    N_B ← 0;    p' ← p;
16.                        FOR all x ∈ p' DO    // compute
```

© Springer International Publishing Switzerland 2016
S. Rehman et al., *Reliable Software for Unreliable Hardware*,
DOI 10.1007/978-3-319-25772-3

```
number of consecutive instructions of Type B
17.                               IF (x == typeB) THEN
18.                               N_B ← N_B + 1;      p'.remove(x);
19.                               ELSE
20.                               NB ← 0;             p'.remove(x);
21.                          END IF
22.                     END FOR
23.                 IF (N_B < 1) THEN            // compute mask-
ing probabilities
24.                     P_DP (I, p) ← P_D (I) + (1 - P_D (I)) ×
P_D (I.s);
25.                 ELSE
26.                     P_D' (I) ← Σ P_D(x);
27.                     P_DP (I, p) ← P_D' (I) + (1 - P_D' (I))
× P_D (I.s);
28.                 END IF
29.            END FOR
30.            L.remove(I);
31.            FOR all ip ∈ I.P DO
32.                 L.add(ip);
33.            END FOR
34.         END FOR
35.    END WHILE
```

Equation at line 26:

$$P_D' (I) \leftarrow \sum_{x=1}^{1+N_B} P_D(x);$$

C.2: Algorithm for Computing the Instruction Error Propagation Index

```
Instruction Error Propagation Index Computation

Input: Instruction flow graph G (V, E), set of leaf nodes L_G,
masking probabilities due to dataflow P_DP(I, p),control flow
probabilities P_CF (p | I).
Output: set of error propagation indices for each instruction
I, EPI (I).
1.    FOR all I ∈ L_G(G) DO
2.         EPI (I) ← 1; // initialization to consider error
propagation of leaf nodes
3.    END FOR
4.    List C(L_G);              Queue Q();// list of traversed
```

```
instructions and queue of evaluated instructions
5.      FOR all I ∈ L_G(G) DO
6.            FOR all i ∈ I.P DO
7.                  Q.Enqueue(i);
8.            END FOR
9     END FOR
10.     WHILE (!Q.isEmpty()) DO
11.              I ← Q.Dequeue(I);
12.              IF (∀ s ∈ I.S, s ∈ C) THEN     //  compute
EPI for all instructions with successors in C
13.              EPI ← 0;
14.                FOR all s ∈ I.S DO
15.                IF (IMI(s) == 0) THEN    // if successor
is a non-masking instruction
16.                      EPI ← EPI + (EPI(s) × P_{Execution}(s |
I));
17.                    ELSE
18.                      EPI ← EPI +
```

$$\sum_{\forall p \in s.Paths} \left(\left(1 - P_{DP}\left(s,p\right)\right) \right) \times P_{CF}\left(p \mid I\right) \times EPI\left(L_G\left(p\right)\right) \right) \quad ;$$

```
19.                    END IF
20.              END FOR
21.                EPI(I) ← EPI × (P_{IO}(I) / (P_{IO}(I) +
P_{Cr}(I)));      C.add(I);
22.              FOR all I ∈ I.P DO
23.                  Q.Enqueue(I);
24.              END FOR
25.          ELSE
26.              Q.Enqueue(I);
27.          END IF
28.     END WHILE
```

C.3: Algorithm for FVI-Driven Data Type Optimization

Algorithm C.3 presents the pseudo-code targeting load merging (for store instructions, the procedure is similar).

Input: Graph $G(V, E)$ of the function F, P_τ as the tolerable performance overhead, Data Type, *FVI* and performance of the original code (FVI_{Orig}, P_{Orig}).

Output: Transformed function *fd* with merged loads and extraction code as a result of the data type optimization.

```
FVI-Driven Data Type Optimization

Input: G (V, E), Pₜ, FVI_orig, P_orig, DataType
Output: Transformed function fd
1.         A ← getAllArrays(G);
2.         FOR all a ∈ A DO
3.             List <V> L ← getLoads(a, G);
4.             IF (DataType == INT) THEN
5.                 continue;
6.             END IF
7.             FVI_Best ←FVI_orig;
8.                 WHILE L ! = Ø DO
9.                     G' ← G;
10.                    (l₁, l₂) ← getCurrent&NextLoads(L);
11.                    l ← Merge(l₁, l₂);
12.                       G'.remove(l₁, l₂);    G'.insert(l);    G
'.insertExtractionCode();
13.                       (FVI, P, Spill) ← Evaluate(G'); // compile
and execute, estimate FVI, performance, and
                                                          check
for spilling
14.                    IF ((P/P_Orig - 1) > Pₜ) THEN
15.                        break;
16.                    END IF
17.                       IF ((FVI < FVI_Best) && (!Spill)) THEN
18.                           FVI_Best ← FVI;    L.remove(l₁,l₂);
19.                        G.remove(l₁,l₂);    G.insert(l);
20.                           G.insertExtractionCode();
21.                    END IF
22.                 END WHILE
23.        END FOR
24.        fd ← G;
25.        return fd;
```

C.4: Algorithm for FVI-Driven Loop Unrolling

Input: a set of maximum unrolling factors for all loops, the FVI, performance, and code size of the original function *F*.

Output: the transformed function *fd* with loop unrolling applied by an *FVI-minimizing unrolling factor*.

FVI-Driven Loop Unrolling

Input: Function F, Set of maxUnrollFactors, FVI_{Orig}, P_{Orig}, C_{Orig}, μ.

Output: Transformed function fd

1.   list< Loop > L ← getLoops(F);

2.   FOR all l ∈ L DO // determine unrolling factor for each loop

3.   maxUF = getFactor(l, maxUnrollFactors);

4. $unrollProfit_{Best}$ ← minINT; uF_{Best} ← 1;

5.   FOR uF_i = 1 to maxUF DO

6. l_{temp} ← l;   // create a temporary copy of the loop

7. F_{uFi} ← Unroll (F, l_{temp}, uF_i);   // Unroll by a factor uF_i

8.   (FVI, P, C, Spill) ← Evaluate(F_{uFi}); // compile. execute, estimate FVI and
                                          performance, and check for spilling.

9. $FVI_{Benefit}$ ← (FVI_{Orig} − FVI)/FVI_{Orig};

10. P_{Loss} ← (P − P_{Orig})/P_{Orig}; // Performance loss

11. C_{Loss} ← (C − C_{Orig})/C_{Orig}; // Code size increase

12.   unrollProfit = computeProfit($FVI_{Benefit}$, P_{Loss}, C_{Loss}, μ);

13.   IF ((unrollProfit > $unrollProfit_{Best}$) && (!Spill)) THEN

14. $unrollProfit_{Best}$ ← unrollProfit; uF_{Best} ← uF_i;

15.   END IF

16.   END FOR

17.   setBestUnrollFactor(l, uF_{Best});

18.   END FOR

19.   FOR all l ∈ L DO // generate the transformed function using the best unroll factors

20. UF_{Best} = getBestUnrollFactor(l);

21.   fd ← Unroll(fd, l, UF_{Best});

22.   END FOR

23.   return fd;

C.5: Algorithm for Applying Common Expression Elimination

Algorithm C.5 shows the pseudo-code of the algorithm to evaluate the reliability benefit of replacing common expressions.

Input: Graph G (V, E) of the function F, P_τ as the tolerable performance overhead, FVI and performance of the original code (FVI_{Orig}, P_{Orig}).

Output: Transformed function *fd* where the common expressions are (partially) replaced.

```
Common Expression Elimination

Input: G (V, E), Pᵣ, FVIₒᵣᵢ_g, Pₒᵣᵢg
Output: Transformed function fd
1.    CG ← getCEs(G); // get all common sub-graphs/expressions
in G
2.     FOR all c ∈ CG DO //  Evaluation Phase
3.           O ← getOccurrences(c, G);
4.         FOR all o  ∈  O DO
5.              f₁ ← computeFVI(replace CE at o in G);
6.              f₂ ← computeFVI(keep CE at o in G);
7.              Δ ←f₂ - f₁;
8.              IF (Δ ≥ 0) THEN
9.                   o.mode ← set(replace_CE);
10.             ELSE
11.                  o.mode ← set(keep_CE);
12.             END IF
13.          END FOR
14.          G' ← updateGraph(G); //  replace occurrences
based on o.mode
15.              FVIc ← computeFVI(G'); //  FVI
16.              Pc ← computePerformance(G'); //
performance
17.              εc ← (FVIₒᵣᵢg - FVIc)/(Pc - P_Best); //
efficiency
18.    END FOR
19.    sort(L_CE); //  sort common sub-graphs/expression by
their efficiency
20.    FOR all l ∈ L_CE DO //  Elimination Phase
21.        IF (l.Pc - P_Best ≤ Pᵣ) THEN
22.            O ← getOccurences(l.c, G);
23.            FOR all o ∈ O DO
24.                IF ((o.mode == replace_CE) && (!Spill))
```

```
THEN
25.                      G.remove(o);    G.insert(CE_variable);
26.                         Pᵣ ← Pᵣ - o.latency();
27.                   END IF
28.               END FOR
29.            END IF
30.      END FOR
31.      fd ← G
32.      return fd; //   return the code with expression
elimination
```

C.6: Algorithm for Soft-Error-Driven Instruction Scheduler

```
Soft Error Driven Instruction Scheduler
```

```
Lookahead (): Input: Instruction Graph G = (V, E), Tolerable
performance overhead Pᵣ.
         Output: Instruction Schedule Gₛ.
33.     Gₛ ← ∅; G_SC ← ∅; //   set of scheduled and candidate
instructions
34.     FOR all n ∈ V DO
35.         S[n] ← n.getSucc();    P[n] ← n.getPred();
36.         IF (ready(n)) THEN
37.             G_SC ← G_SC ∪ n; //   add to the ready list
38.         END IF
39.     END FOR
40.     FOR all n ∈ G_SC DO
41.         eT[n] ← 0; //   initialization of the earliest time
42.     END FOR
43.     T_curr ← 0;    i_PSel ← ∅ ;    i_RSel ← ∅ ; //   Performance
and reliability maximizing instruction
44.     Ψ_max ← - ∞;    δ_max ← - ∞
45.     WHILE (G_SC != ∅ ) DO
46.         FOR all i ∈ G_SC DO
47.                 Ψ[i] ← estimateReliabilityWeight(i); //
Eqs. 5.2-5.5;
48.                 IF ((δ[i] > δ_max) && eT[i] ≤ T_curr) THEN //
obtain instruction considering performance
49.                     δ_max ← δ[i];        i_PSel ← i ;
50.                 END IF
```

```
51.             IF (P_τ > 0) THEN
52.                 FOR all j ∈ (G_SC - {i}+ j.getSchedulableSN())
DO//  evaluate candidates with lookahead
53.                     Ψ_ij ← Ψ[i] + Ψ[j];          δ_Loss ←
δ_max - δ[i];
54.                     IF ((Ψ_ij > Ψ_max) && (δ_Loss < P_τ)) THEN
55.                         Ψ_max ← Ψ_ij;       i_RSel ← i;
56.                     ELSE IF ((Ψ_ij == Ψ_max) && (δ_Loss < P_τ) &&
(δ[i] > δ[i_RSel]) && (eT[i] ≤ Tcurr)) THEN
57.                         Ψ_max ← Ψ_ij;       i_RSel ← i;
58.                     END IF
59.                 END FOR
60.                 IF (i_RSel != ∅ ) THEN
61.                     i_Sel ← i_RSel;        P_τ ← P_τ - (δ_max - δ[i_Sel]);
//  select reliability-wise best solution
62.                 ELSE
63.                     i_Sel ← i_PSel; //  select performance-wise
best solution
64.                 END IF
65.             ELSE
66.                 i_Sel ← i_PSel;
67.             END IF
68.         END FOR
69.         T_curr ← T_curr + T[i_Sel];     G_SC← G_SC - i_Sel;        G_S ←
G_S ∪ i_Sel;
70.         FOR all s ∈ S[i_Sel] DO//  update the set of schedul-
ing candidates
71.             IF (∀m ∈ P[s] ∃ t | V_GS[t] = m) THEN
72.                 G_SC ← G_SC ∪ s;    eT[s] ← T_curr + eT[i_Sel];
73.             END IF
74.         END FOR
75.     END WHILE
76.     return G_S;
```

C.7: Algorithm for Selective Instruction Protection Technique

```
Selective Instruction Protection
─────────────────────────────────────────────────────

Input: Unprotected function F from the software program as G =
(V, E), user provided tolerable performance overhead in cycles
P_τ, set of instruction vulnerabilities IVI, set of error
```

```
propagation indices for all instructions EPI, user provided
program reliability method R (for instance, SWIFT-R 0)
Output: Function with selective instruction protection F'.
1.    List L;
2.    FOR all i ∈ G DO//  compute the reliability profit for
all instructions
3.        RPF(i) ← (EPI(i) ∈ IVI(i))/ω(i);
4.    END FOR
5.    Sort(L, RPF, Descending order);
6.    WHILE (!L.empty() && (P_r > 0)) DO
7.        I ← L.pull();
8.        IF (ω(I) ≤ P_r ) THEN
9.                Protect(I);        P_r ← P_r - ω(I);
10.               FOR all i ∈ (I.S ∈ I.P) DO
11.                   GI ← generateGroups(I, i); //  groups
of consecutive instructions
12.                   FOR all g ∈ GI DO //  compute overhead of
instruction groups
13.                       g' ← g;
14.                       FOR all i ∈ g DO
15.                           IF ((i.S > I) && (∃_s∈_{I.s}
inGroup(s,g') == False)) THEN
16.                               setCheckInstructionPt(i, g');
17.                           END IF
18.                           setCheckInstructionPt(Leaf(g),
g');
19.                       END FOR
20.                       ω(i, g') ← getOverload(g', R);
21.                   END FOR
22.                   RPF(i) ← (EPI(i) × IVI(i))/ω(i, g');
23.               END FOR
24.           Sort(L, RPF, DescendingOrder);
25.       END IF
26.   END WHILE //  end while loop if budget is over or all
instructions are protected
```

C.8: Algorithm for Offline Table Construction

Algorithm C.8 describes the procedure of offline table construction by adopting the approximations explained in Sect. 6.1, and by using δ as the timing unit and α as the reliability penalty unit.

Offline Table Construction

Input: n functions, CDF and PDF of the functions, units δ and
σ, weighted parameter α, and the default versions θ () after
observing the deadline misses;

1.   FOR r ← $\dfrac{R_{min}(n)}{\sigma}\sigma,...,\dfrac{R_{max}(n)}{\sigma}\sigma$, stepped by σ DO

2.   FOR t ← $0,...,\dfrac{D}{\delta}\delta$, stepped by δ DO

3.   Calculate j*(n, r, t) and G(n, r, t) by using
Eq. 6.2

4.   END FOR

5.   END FOR

6.   FOR i ← n-1, n-2,..., 2 DO

7.   FOR r ← $\dfrac{R_{min}(i)}{\sigma}\sigma,...,\dfrac{R_{max}(i)}{\sigma}\sigma$, stepped by σ DO

8.   FOR t ← $0,...,\dfrac{D}{\delta}\delta$, stepped by δ DO

9.   IF (t == 0) THEN

10.   j*(i, r, t) ← θ$_1$; G(i, r, t) ←
α (r + ρ(i)) + (1 - α)

11.   ELSE

12.   FOR each j = 1, 2,..., K$_i$,

13.   Calculate H$_j$ ←
$$\int_{x=0}^{t} P_{i,j}(x).G\left(i+1,\frac{r+R_{i,j}}{\sigma}\sigma,\frac{t-x}{\delta}\delta\right)dx$$ +
$$(1-C_{i,j}(t)).(\alpha(r+R_{i,j}+\rho(i+1))+(1-\alpha));$$

14.   j*(i, r, t) ←argmin j=1,2,....,Ki
H$_j$;

15.   G(i, r, t) ← $H_{j^*(i,r,t)}$

16.   END IF

17.   END FOR

18.   END FOR

19.   END FOR

20.   Calculate j*(1, 0, D) and G (1, 0, D) using the same
procedure as in Steps 14 and 15.

21.   Return the table j*.

C.9: Algorithm for Hybrid RMT Tuning

Hybrid RMT Tuning

Input: set T of tasks; a list of available/free cores List$_c$; list of different compiled versions {t$_1$ = {t$(_{1,1})$,…, t$(_{1,K1})$}},…, t$_M$ = {t$(_{M,1})$, …, t$(_{M,KM})$}; history H of last S$_H$ RMT comparison results; list TR{$_{RMT,NR}$} of running tasks (sorted by RMT level and RTP).

Output: performance-wise sorted version lists V$_{t1}$,…, V$_{tM}$ containing different reliability performance tradeoffs per task, list L of tasks with a protection type pt and number of allocated cores rc used for RMT.

1.   FOR all t ∈ T DO // estimate performance and RTP for all task versions

2.   FOR all cv ∈ t DO   // loop over all task versions

3.   cv.L = estimatePerformance(cv);

4.   cv.RTP = calculateRTpenalty(cv);

5.   V$_t$.insert(cv);

6.   END FOR

7.   V$_t$.sort(cv.L);

8.   END FOR

9.   List TL;

10.   FOR all t ∈ T DO

11.   t.v = V$_t$.head(1); TL.insert(t);

12.   END FOR

13.   TL.sort(t.RTP);   // sort task list by RTP

14.   N$_{FC}$ = List$_c$.size(); // number of free cores

15.   IF (TR$_{RMT}$.size() == 0 && N$_{FC}$ > 1) THEN // at least one task with RMT

16.   t = TL.pop_front(); t.pt = RMT; TR$_{RMT}$.push_b ack(t);

17.   rc = min (coreDemandRMT, N$_{FC}$); N$_{FC}$ = N$_{FC}$ - rc; t.rc = rc;

18.   END IF

19.   WHILE (N$_{FC}$ > 0 && TL.size() > 0) DO // allocate a core to each task 20.   t = TL.pop_front(); t.pt = NR; T$_{NR}$.push_back(t);

21.   N$_{FC}$ -- ; t.rc = 1;

```
22.    END WHILE
23.    WHILE (TL.size() > 0) DO // preemption for tasks run-
ning with RMT
24.              IF (TR_RMT.size() > 1) THEN
25.                   t_r = TR_RMT.back();
26.              ELSE
27.                   break;
28.              END IF
29.              IF (t_r.rc == 2) THEN    // 2 is the required
number of cores for RMT
30.                   t_r.pt = NR; T_NR.push_front(t_r);
TR_RMT.pop_back();
31.              END IF
32.              t_r.rc = t_r.rc - 1;
33.         t = TL.pop_front(); t.pt = NR; TR_NR.push_back(t);
t.rc =1;
34.    END WHILE
```

35. $h = \sum_{i=1}^{5} H[i]*i$; // evaluate RMT comparison history

```
36.    IF (h> SH) THEN // activate RMT mode based on comparison
history
37.         WHILE (N_FC ≥ coreDemandRMT - 1) DO
38.         t = T_NR.pop_front(); t.pt = RMT; t = TR_RMT.insert();
39.              N_FC = N_FC - coreDemandRMT - 1; t.rc =
coreDemandRMT;
40.         END WHILE
41.    END IF
```

C.10: Algorithm for Reliable Code Version Tuning

```
Version Tuning

Input: lists T_pt of tasks with protection levels pt ∈ {NR,
RMT}; performance-wise sorted version lists V_t1,...,V_tM; list of
available/free cores List_c.
Output: tasks T with set of selected versions {t(_i,_j), ∀ i ∈
T, j ∈ K_i }
1.    FOR all t ∈ T_RMT ∪ T_NR DO
2.         t.c = List_c.tail(1);
3.    END FOR
```

```
4.      FOR all t ∈ T_RMT ∪ T_NR DO  // loop over all tasks
5.         v = V_t.head(1);
6.          FOR all j ∈ {1,…,t.rc} DO        // initialize with
performance-wise best version
7.             t.v_j = v;
8.          END FOR
9.             IF (t.rc ≤ 2) THEN   // tasks running with no
redundancy
10.              FOR all v ∈ V_t  DO
11.                  IF (v.RT > t.v_1.RT && v.L < t.D) THEN  //
select the version with best reliability within the deadline
constraint
12.                      FOR all j ∈ {1,…,t.rc} DO
13.                          t.v_j = v;
14.                      END FOR
15.                   END IF
16.               END FOR
17.          END IF
18.      END;
19.      FOR all t ∈ T_RMT ∪ T_NR DO
20.          t.c =  ∅;
21     END FOR
```

C.11: Algorithm for Core Tuning and Version Update

Algorithm C.11: Pseudo-Code for the Core Tuning and Version Update

```
Core Tuning and Final Version Tuning
```

```
Input: task graph G, lists T_pt of tasks with protection levels
pt ∈ {NR, RMT}; performance-wise sorted version lists V_t1,…
,V_tM; performance-wise sorted list of available/free cores
List_c.
Output: tasks T with set of selected versions {t(_i,_j), ∀ i ∈
T, j ∈ K_i } and set of assigned cores {t.c_1,…,t.c_t.rc}
1.       List L = T_RMT ∪ T_NR;     L.sort(t.RTP);
2.        FOR all t ∈ L DO //  loop over all tasks
3.            List L_dC = t.getDependentTasks(G); //  list of
dependent cores
```

```
4.              List L_exec;
5.               FOR all c ∈ List_c.head(3) DO //  analyze 3 per-
formance-wise best cores
6.                  comm = 0;
7.                      FOR all t_dep ∈ L_dC DO //  analyze com-
munication overhead
8.                          dist = computeDistance(c, t_dep.c);
9.                          comm += estimateCommOverhead(dist,
t_dep.data);
10.                         perf = estimatePerformance(c,t.v);
11.                     END FOR
12.                 t.exec = comm + perf;
13.                 L_exec.insert(pair< c,t.exec >);
14.             END FOR
15.                 t.c_1 = getCore(min(∀ t.exec ∈ L_exec)); //
select core with lowest combination of execution time  and
communication
16.             List_c.remove(t.c_1);
17.          i = 2;
18.              WHILE (i ≤ t.rc) DO // find close cores for
redundant executions
19.                  t.c_i = min (∈ c ∈ List_c {calcDistance(t.
c_1,c)});
20.                  List_c.remove(t.c_i); i + +;
21.          END WHILE
22.      END FOR
23.      FOR all t ∈ L DO
24.          tuneVersion(t);
25.      END FOR
```

Appendix D
Notations and Symbols

This appendix presents the table of notations/symbols/terms and their descriptions used in this manuscript. The parameters/symbols are listed in the alphabetic order in the following table.

Parameter/symbol	Description
A_c	Area of the processor component c
BB	Basic Block
CAB(I)	Number of critical address bits of the instruction I that lead to "memory segmentation" errors
ci	Checking instructions
CI	Critical Instruction
COB(I)	Number of critical operand bits of the instruction I that lead to "non-decodable instruction" errors
$C_{i,j}(D)$	Probability for which the execution time of jth version of function i is less than or equal to the deadline D
C_{Orig}	Code size of the of the Original Code without reliability-driven transformations
csi	Consecutive instructions
D	Application Deadline
DB	Dependent Basic Blocks
E	Set of edges in an application graph G
ep	Execution Path
EPI	Instruction Error Propagation Index
ER_{raw}	Raw error rate
f	Fault rate
FVI	Function Vulnerability Index
$FVI_{Failures}$	Function's vulnerability to Application Failures
$FVI_{IncorrectOP}$	Function's vulnerability to Incorrect Outputs

(continued)

© Springer International Publishing Switzerland 2016
S. Rehman et al., *Reliable Software for Unreliable Hardware*,
DOI 10.1007/978-3-319-25772-3

Parameter/symbol	Description
FVI_{Orig}	FVI of the of the Original Code without reliability-driven transformations
G	Application graph with a set of vertices V (or T as a set of Tasks) and edges E
$G^*(i, r, t)$	A single function version entry in the table of function schedules corresponding to the instruction i, with reliability level r and time t
$I.Paths$	A set of paths p for an instruction I
I_D	Dependent instruction
IMI	Instruction Error Masking Index
$IMI(I)$	Instruction Masking Index of instruction I
i_{PSel}	Performance-maximizing selected instruction during instruction scheduling
i_{RSel}	Reliability-maximizing (i.e., vulnerability minimizing) selected instruction during instruction scheduling
IVI	Instruction Vulnerability Index
IVI_i	Instruction Vulnerability Index of instruction i
IVI_{ic}	Instruction Vulnerability Index of instruction i in processor component c
$j^*(i, r, t)$	Index of the function version entry in the table of function schedules corresponding to the instruction i, with reliability level r and time t
$maxUnrollFactor$	Maximum value of the unrolling factor for a given function
$miss\ rate$	Percentage of deadline misses for a given application
nCI	Non-critical Instruction
P	A set of predecessor instructions for a given instruction
p	An instruction path, such that each instruction in the path has exactly one successor and one predecessor instruction
PC	Set of different pipeline stages $PC = \{F, D, E, M, W\}$, where F = Instruction Fetch, D = Instruction Decoder, E = Execute, M = Memory, W = Writeback stage in a 5-stage pipeline processor like LEON2
Proc	Set of processor components, i.e., $c \in$ Proc
$P_{AF}(I)$	Probability of Application Failures in case a fault occurs during the execution of the instruction I
$P_{CF}(ep\|I)$	Execution Path Probability for an instruction I given an execution path ep
$P_D(I)$	Error Masking Probability during the execution of an instruction I with O as a set of operands
$P_D(x, I)$	Error Masking Probability for the operand bit x during the execution of instruction I
$P_{DP}(I, p)$	Error Masking Probability, due to data flow properties, for an instruction I along the path p until the final visible program output at the end of the path
$P_e(x)$	Error Probability in an operand bit x
$P_{eAd}(b, I)$	Error Probability in the address bits b of the instruction I
$P_{EM}(c)$	Error Masking Probability of a processor component c

(continued)

Parameter/symbol	Description
$P_{\text{EM}}(i, PC)$	Error Masking Probability for an instruction i in the pipeline stage PC
$P_{\text{eOP}}(b, I)$	Error Probability in the opcode bits b of the instruction I
$P_{\text{Execution}}(s\|I)$	Execution probability of a successor instruction s corresponding to an instruction I
P_{Failures}	Probability of Application Failures
$P_{\text{fault}}(c)$	Probability of Fault in a processor component c
$P_{\text{IncorrectOP-CI}}$	Probability of Incorrect Outputs due to critical instructions
$P_{\text{IncorrectOP-nCI}}$	Probability of Incorrect Outputs due to non-critical instructions
$P_{\text{IO}}(I)$	Probability of Incorrect Outputs in case a fault occurs during the execution of the instruction I
P_{Orig}	Performance (in terms of execution time given as *cycles*) of the Original Code without reliability-driven transformations
P_{sig}	Signal probability
$P\tau$	Tolerable performance overhead
Q	Queue
R	Reliability function for quantifying the functional correctness. It can either be FVI or function resilience or any other function reliability metric
R_{max}	Maximum range of the RTP value for a given function
R_{min}	Minimum range of the RTP value for a given function
RTP	Reliability-Timing Penalty
S	A set of successor instructions for a given instruction
$TotalBits_c$	Total Bits representing the architecturally defined size of the processor component c
ω	Protection overhead
$vulBits_{ic}$	Vulnerable Bits of the processor component c executing instruction i
$vulP_{ic}$	Vulnerable Period of instruction i in processor component c
V	Set of vertices in an application graph G
ψ	Instruction Reliability Weight as a joint function of IVI, criticality and dependent instructions
ψF	Function Reliability Weight of a function F
$\Pi_{i,j}$	Probability of deadline misses for a given function version j of a function i

Bibliogrphy

1. G. Moore, "Cramming more components onto integrated circuits", *Electronics*, vol. 38, no. 8, 1965.
2. Intel, http://www.intel.com/pressroom/kits/quickref.htm [Online; accessed: Apr 2015].
3. Intel, http://ark.intel.com/products/80555/Intel-Xeon-Phi-Coprocessor-7120A-16GB-1_238-GHz-61-core [Online; accessed Apr 2015].
4. Nvidia, http://www.nvidia.com/object/white-papers.html [Online; accessed Apr 2015].
5. International Technology Roadmap for Semiconductors, in *ITRS 2013 Edition—Process Integration, Devices, and Structures*, 2014.
6. K. Bernstein, D. J. Frank, A. E. Gattiker, W. Haensch, B. L. Ji, S. R. Nassif, E. J. Nowak, D. J. Pearson, and N. J. Rohrer, "High-performance CMOS variability in the 65-nm regime and beyond", *IBM Journal of Research and Development—Advanced silicon technology*, vol. 50, pp. 433–449, Jul. 2006.
7. S. Mitra, K. Brelsford, Y. Kim, H. Lee, and Y. Li, "Robust System Design to Overcome CMOS Reliability Challenges", *IEEE Journal on Emerging and Selected Topics in Circuits and Systems*, vol. 1, no. 1, pp. 30–41, 2011.
8. A. W. Strong, E. Y. Wu, R. P. Vollertsen, J. Sune, G. La Rosa, T. D. Sullivan, and S. E. Rauch III, "Reliability Wearout Mechanisms in Advanced CMOS Technologies", *Wiley-IEEE Press*, vol. 12, ISBN: 978-0471731726, 2009.
9. S. Borkar and A. A. Chien, "The Future of Microprocessors", *Communications of the ACM*, vol. 54, no. 5, pp. 67–77, 2011.
10. S. Borkar, "Designing Reliable Systems from Unreliable Components: The Challenges of Transistor Variability and Degradation", *IEEE Micro*, vol. 25, no. 6, pp. 10–16, 2005.
11. A. Leon, B. Langley, and J. Shin, "The UltraSPARC T1 processor: CMT reliability", in *Proceedings of the Custom Integrated Circuits Conference*, pp. 555–562, 2006.
12. T. Austin, V. Bertacco, S. Mahlke, and Y. Cao, "Reliable Systems on Unreliable Fabrics", *IEEE Transactions on Design & Test of Computers*, vol. 25, no. 4, pp. 322–332, 2008.
13. M. A. Alam, S. Mahapatra, "A comprehensive model for PMOS NBTI degradation", *Microelectronics Reliability*, pp. 71–81, 2005.
14. S. Borkar, T. Karnik, S. Narendra, J. Tschanz, A. Keshavarzi, and V. De., "Parameter variations and impact on circuits and microarchitecture", in *Proceedings of the 40th Annual Design Automation Conference (DAC)*, pp. 338–342, ACM, 2003.
15. K. Bowman, A. Alameldeen, S. Srinivasan, and C. Wilkerson. "Impact of die-to-die and within-die parameter variations on the clock frequency and throughput of multi-core processors", *IEEE Transactions on Very Large Scale Integration (VLSI) Systems,* vol. 17, no. 12, pp. 1679–1690, 2009.

© Springer International Publishing Switzerland 2016
S. Rehman et al., *Reliable Software for Unreliable Hardware*,
DOI 10.1007/978-3-319-25772-3

16. R. Rajeev, A. Devgan, D. Blaauw, and D. Sylvester, "Parametric yield estimation considering leakage variability", in *Proceedings of the 41st Annual Design Automation Conference (DAC)*, pp. 442–447, ACM, 2004.

17. D. Cheng and S. K. Gupta, "Maximizing yield per area of highly parallel cmps using hardware redundancy", *IEEE Transactions on Computer-Aided Design of Integrated Circuits and Systems (TCAD)*, vol. 33, no. 10, pp. 1545–1558, Oct 2014.

18. R. Baumann, "Radiation-induced soft errors in advanced semiconductor technologies", *IEEE Transactions on Device and Materials Reliability,* vol. 5, no. 3, pp. 305–316, 2005.

19. J. Henkel, L. Bauer, N. Dutt, P. Gupta, S. Nassif, M. Shafique, M.Tahoori, and N.Wehn, "Reliable on-chip systems in the nano-era: Lessons learnt and future trends", in *Proceedings of the 50th Annual Design Automation Conference (DAC)*, pp. 99, ACM, 2013.

20. "Firmware-based Platform Reliability", Intel Corporation, 2004.

21. P. Shivakumar, M. Kistler, S. Keckler, D. Burger, and L. Alvisi, "Modeling the effect of technology trends on the soft error rate of combinational logic", in *Proceedings of the IEEE International Conference on Dependable Systems and Networks (DSN)*, pp. 389–398, 2002.

22. S. Mukherjee., J. Emer, and S. Reinhardt, "The soft error problem: An architectural perspective", in *The 11th International Symposium on High-Performance Computer Architecture, 2005. HPCA-11*, pp. 243–247, 2005.

23. R. Lefurgy, A. Drake, M. Floyd, M. Allen-Ware, B. Brock, J. Tierno, and J. Carter, "Active management of timing guardband to save energy in POWER7", in *Proceedings of the 44th Annual IEEE/ACM International Symposium on Microarchitecture*, pp. 1–11, ACM, 2011.

24. M. Agarwal, B. C. Paul, M. Zhang, and S. Mitra, "Circuit Failure Prediction and Its Application to Transistor Aging", in *Proceedings of the VLSI Test Symposium*, pp. 277–286, 2007.

25. B. Raghunathan, Y. Turakhia, S. Garg, and D. Marculescu, "Cherry-picking: exploiting process variations in dark-silicon homogeneous chip multi-processors", in *Proceedings of the Conference on Design, Automation and Test in Europe (DATE)*, pp. 39–44. EDA Consortium, 2013.

26. P. Gupta, Y. Agarwal, L. Dolecek, N. Dutt, R. Gupta, R. Kumar, S. Mitra, A. Nicolau, T.Rosing, M. Srivastava, S. Swanson, and D. Sylvester, "Underdesigned and opportunistic computing in presence of hardware variability", in *IEEE Transactions on Computer-Aided Design of Integrated Circuits and Systems (TCAD)*, vol. 32, no. 1, pp. 8–23, 2013.

27. N. Oh, P. Shirvani, and E. McCluskey, "Error detection by duplicated instructions in superscalar processors", in *IEEE Transactions on Reliability*, vol. 51, no. 1, pp. 63–75, 2002.

28. G. Reis, J. Chang, N. Vachharajani, R. Rangan, D. August, and S. Mukherjee, "Software-controlled fault tolerance", in *ACM Transactions on Architecture and Code Optimization (TACO)*, vol. 2, no. 4, pp. 366–396, 2005.

29. S. Borkar, "Microarchitecture and design challenges for gigascale integration", in *MICRO*, vol. 37, p. 3, 2004.

30. E. Ibe, H. Taniguchi, Y. Yahagi, K. Shimbo, and T. Toba, "Impact of Scaling on Neutron-Induced Soft Error in SRAMs From a 250 nm to a 22 nm Design Rule", in *IEEE Transactions on Electron Devices*, vol. 57, no. 7, pp. 1527–1538, 2010.

31. S. Mukherjee, C. Weaver, J. Emer, S. Reinhardt, and T. Austin, "A systematic methodology to compute the architectural vulnerability factors for a high-performance microprocessor", in *Proceedings of the 36th Annual IEEE/ACM International Symposium on Microarchitecture (MICRO)*. pp. 29, 2003.

32. N. Seifert, B. Gill, S. Jahinuzzaman, J. Basile, V. Ambrose, Q. Shi, R. Allmon, and A. Bramnik, "Soft error susceptibilities of 22 nm tri-gate devices", in *IEEE Transactions on Nuclear Science,* vol. 59, no. 6, pp. 2666–2673, 2012.

33. R. Vadlamani, J. Zhao, W. Burleson, and R. Tessier, "Multicore soft error rate stabilization using adaptive dual modular redundancy", in *IEEE Design, Automation and Test in Europe Conference & Exhibition (DATE)*,pp. 27–32, 2010.

34. N. Oh, P. Shirvani, and E. McCluskey, "Control-flow checking by software signatures", in *IEEE Transactions on Reliability,* vol. 51, no. 1, pp. 111–122, 2002.

35. J. Gaisler, "A portable and fault-tolerant microprocessor based on the SPARC v8 architecture", in *Proceedings of the IEEE/IFIP International Conference on Dependable Systems and Networks(DSN)*, pp. 409–415, 2002.

36. S. Mukherjee, M. Kontz, and S. Reinhardt, "Detailed design and evaluation of redundant multi-threading alternatives", in *Proceedings of the 29th Annual IEEE International Symposium on Computer Architecture (ISCA)*, pp. 99–110, 2002.

37. A. Shye, J. Blomstedt, T. Moseley, V. Reddi, and D. Connors, "PLR: A software approach to transient fault tolerance for multicore architectures", in *IEEE Transactions on Dependable and Secure Computing*, vol. 6, no. 2, pp. 135–148, 2009.

38. J. Smolens, B. Gold, B. Falsafi, and J. Hoe, "Reunion: Complexity-effective multicore redundancy", in *Proceedings of the 39th Annual IEEE/ACM International Symposium on Microarchitecture (MICRO)*, IEEE Computer Society, pp. 223–234, 2006.

39. C. Constantinescu, "Trends and challenges in VLSI circuit reliability", in *IEEE Micro*, vol. 23, no. 4, pp. 14–19, 2003.

40. H. Kufluoglu and M. Alam, "A Generalized Reaction–Diffusion Model With Explicit H–Dynamics for Negative-Bias Temperature-Instability (NBTI) Degradation", in *IEEE Transactions on Electron Devices*, vol. 54, no. 5, pp. 1101–1107, 2007.

41. H. Hanson, K. Rajamani, J. Rubio, S. Ghiasi, and F. Rawson, "Benchmarking for Power and Performance", in *SPEC Benchmarking Workshop*, 2007.

42. S. Dighe, S. Vangal, P. Aseron, S. Kumar, T. Jacob, K. Bowman, J. Howard, J. Tschanz, V. Erraguntla, N. Borkar, V. De, and S. Borkar, "Within-die variation-aware dynamic-voltage-frequency scaling core mapping and thread hopping for an 80-core processor", in *IEEE International Solid-State Circuits Conference*, 2010.

43. L. Wanner, C. Apte, R. Balani, P. Gupta, and M. Srivastava, "Hardware variability-aware duty cycling for embedded sensors", in *IEEE Transactions on VLSI*, 2012.

44. J. Xiong, V. Zolotov, and L. He, "Robust extraction of spatial correlation", in *IEEE Transactions on Computer Aided Design (TCAD)*, vol. 26, no. 4, pp. 619–631, 2007.

45. S. Herbert and D. Marculescu, "Characterizing chip-multiprocessor variability-tolerance", in *IEEE Design and Automation Conference*, pp. 313–318, 2008.

46. P. Murley and G. Srinivasan, "Soft-error Monte Carlo modeling program, SEMM", in *IBM Journal of Research and Development*, vol. 40, no. 1, 1996.

47. M. Omana, G. Papasso, D. Rossi, and C. Metra, "A Model for Transient Fault Propagation in Combinatorial Logic", in *Proceedings of the 9th IEEE International On-Line Testing Symposium (IOLTS)*, pp. 11–115, 2003.

48. S. Krishnaswamy, G. F. Viamonte, I. L. Markov, and J. P. Hayes, "Accurate Reliability Evaluation and Enhancement via Probabilistic Transfer Matrices", in *Proceedings of Design, Automation and Test in Europe (DATE)*, pp. 282–287, 2005.

49. Y. Dhillon, A. Diril, and A. Chatterjee, "Soft-Error Tolerance Analysis and Optimization of Nanometer Circuits", in *Proceedings of Design, Automation and Test in Europe (DATE)*, pp. 288–293, 2005.

50. S. Kiamehr, M. Ebrahimi, F. Firouzi, and M. Tahoori, "Chip-level modeling and analysis of electrical masking of soft errors", in *The 31st IEEE VLSI Test Symposium (VTS)*, pp. 1–6, 2013.

51. H. Asadi, and M. Tahoori, "An Accurate SER Estimation Method Based on Propagation Probability", in *Proceedings of Design, Automation and Test Conference in Europe (DATE)*, 2005.

52. M. Ebrahimi., L. Chen, H. Asadi, and M. Tahoori, "CLASS: Combined logic and architectural soft error sensitivity analysis", in *18th Asia and South Pacific Design Automation Conference (ASP-DAC)*, pp. 601–607, 2013.

53. K. Itoh, R. Hori, H. Masuda, Y. Kamigaki, H. Kawamoto, and H. Katto, "A single 5V 64k dynamic ram", in *IEEE International Solid-State Circuits Conference (ISSCC)*, Digest of Technical Papers, vol. 23, pp 228–229, 1980.

54. M. Kohara, Y. Mashiko, K. Nakasaki, and M. Nunoshita, "Mechanism of electromigration in ceramic packages induced by chip-coating polyimide", in *IEEE Transactions on Components, Hybrids, and Manufacturing Technology*, vol. 13, no. 4, pp. 873–878, 1990.

55. M. Bruel, "Silicon on insulator material technology", in *Electronics Letters*, vol. 31, no. 14, pp. 1201–1202, 1995.

56. E. Cannon, D. Reinhardt, M. Gordon, and P. Makowenskyj, "Sram ser in 90, 130 and 180 nm bulk and soi technologies", in *Proceedings of 42nd Annual IEEE International Reliability Physics Symposium*, pp. 300–304, 2004.

57. D. Burnett, C. Lage, and A. Bormann, "Soft-error-rate improvement in advanced bicmos srams", in *Proceedings of 31st Annual Reliability Physics Symposium,* pp. 156–160, 1993.

58. S Mitra, T. Karnik, N. Seifert, and M. Zhang, "Logic soft errors in sub-65 nm technologies design and cad challenges", in *Proceedings of 42nd Design Automation Conference (DAC),* pp. 2–4, 2005.

59. D. Ernst, S. Das, S. Lee, D. Blaauw, T. Austin, T. Mudge, N. Kim, and K. Flautner, K, "Razor: circuit-level correction of timing errors for low-power operation", in *IEEE Micro*, vol. 24, no. 6, pp. 10–20, 2004.

60. S. Das, C. Tokunaga, S. Pant, M. Wei-Hsiang, S. Kalaiselvan, K. Lai, D. Bull, and D. Blaauw, "RazorII: In situ error detection and correction for PVT and SER tolerance", in *IEEE Journal of Solid-State Circuits,* vol. 44, no. 1, pp. 32–48, 2009.

61. H. Wunderlich and M. Tahoori, "Tutorial Workshop in the frame of the DFG SPP 1500: Defects, Faults, and Errors - Approaches to Cross-Layer Fault-Tolerance", in *GMM/GI/ITG-Fachtagung Zuverlässigkeit und Entwurf (ZuE)*, 2011.

62. IBM® XIV® Storage System cache: http://publib.boulder.ibm.com/infocenter/ibmxiv/r2/index.jsp [Online; accessed Apr. 2015].

63. AMD Phenom™ II Processor Product Data Sheet 2010.

64. R. Hamming, "Error detecting and error correcting codes", in *Bell System Technical Journal*, vol. 26, no. 2, pp. 147–160, 1950.

65. K. Kang, S. Gangwal, S. Park, and A. Roy, "NBTI Induced Performance Degradation in Logic and 66. Memory Circuits", in *Proceedings of the Asia and South Pacific Design Automation Conference (ASPDAC)*, 2008.

66. Aeroflex, http://aeroflex.com/ams/ [Online; accessed Apr 2015].

67. S. Reinhardt and S. Mukherjee, "Transient Fault Detection via Simultaneous Multithreading", in *Proceedings of the International Symposium on Computer Architecture (ISCA)*, pp. 25–34, 2000.

68. D. Tullsen, S. Eggers, and H. Levy, "Simultaneous multithreading: Maximizing on-chip parallelism", in *ACM SIGARCH Computer Architecture News*, vol. 23, no. 2, pp. 392–403, ACM, 1995.

69. A. Avizienis, "The N-version approach to fault-tolerant software", in *IEEE Transactions on. Software Engineering*, vol. 11, no. 12, pp. 1491–1501, 1985.

70. R. Koo and S. Toueg, "Checkpointing and rollback-recovery for distributed systems", in *IEEE Transactions on Software Engineering*, vol. 1, pp. 23–31, 1987.

71. G. Reis, "*Software modulated fault tolerance*", Ph.D. Thesis, Princeton University, 2008.

72. J. Lee and A.Shrivastava, "A compiler optimization to reduce soft errors in register files", in *ACM Sigplan Notices*, vol. 44, no. 7, pp. 41–49, ACM, 2009.

73. J. Yan and W. Zhang, "Compiler-guided register reliability improvement against soft errors", in *Proceedings of the 5th ACM International Conference on Embedded Software*, pp. 203–209, 2005.

74. V. Sridharan, "Introducing Abstraction to Vulnerability Analysis", *Ph.D. Thesis*, March 2010.

75. V. Sridharan and D. Kaeli, "Eliminating Micro-architectural Dependency from Architectural Vulnerability", in *IEEE International Symposium on High Performance Computer Architecture*, pp. 117–128, 2009.

76. D. Borodin and B. Juurlink, "Protective redundancy overhead reduction using instruction vulnerability factor", in *Proceedings of the 7th ACM International Conference on Computing Frontiers*, pp. 319–326, 2010.

77. J. Hu, S. Wang, and G. Ziavras, "In-register duplication: Exploiting narrow-width value for improving register file reliability", in *IEEE International Conference on Dependable Systems and Networks (DSN 2006)*, pp. 281–290, 2006.

78. P. Lokuciejewski and P. Marwedel, "Combining worst-case timing models, loop unrolling, and static loop analysis for WCET minimization", in *21st IEEE Euromicro Conference on Real-Time Systems (ECRTS)*, pp. 35–44, 2009.

79. V. Sarkar, "Optimized Unrolling of Nested Loops", in *International Journal on Parallel Programing*, vol. 29, no. 5, pp. 545–581, 2001.

80. J. Hu, F. Li, V. Degalahal, M. Kandemir, N. Vijaykrishnan, and M. Irwin, "Compiler-directed instruction duplication for soft error detection", in *Proceedings of the Conference on Design, Automation and Test in Europe (DATE)*, pp. 1056–1057, 2005.

81. J. Xu, Q. Tan, and R. Shen, "The Instruction Scheduling for Soft Errors based on Data Flow Analysis", in *IEEE Pacific Rim International Symposium on Dependable Computing*, pp. 372–378, 2009.

82. L. Spainhower and T. Gregg, "IBM S/390 parallel enterprise server G5 fault tolerance: A historical perspective", in *IBM journal of Research and Development*, vol. 43, no. 5/6, 1999.

83. T. Li, M. Shafique, S. Rehman, J. A. Ambrose, J. Henkel, and S. Parameswaran, "DHASER: Dynamic Heterogeneous Adaptation for Soft-Error Resiliency in ASIP-based Multi-core Systems", in *IEEE International Conference on Computer Aided Design (ICCAD)*, pp. 646–653, 2013.

84. J. Maiz, S. Hareland, K. Zhang, and P. Armstrong, "Characterization of multi-bit soft error events in advanced SRAMs", in *Electron Devices Meeting* (IEDM), pp. 21.4.1–21.4.4, 2003.

85. K. Osada, K. Yamaguchi, Y. Saitoh, and T. Kawahara, "SRAM immunity to cosmic-ray-induced multierrors based on analysis of an induced parasitic bipolar effect", in *IEEE Journal of Solid-State Circuits*, vol. 39, no. 5, pp. 827–833,2004.

86. J.-M. Palau, G. Hubert, K. Coulie, B. Sagnes, M.-C. Calvet, and S. Fourtine, "Device simulation study of the seu sensitivity of srams to internal ion tracks generated by nuclear reactions", in *IEEE Transactions on Nuclear Science*, vol. 48, no. 2, pp. 225–231, 2001.

87. N. Miskov-Zivanov and D. Marculescu, "Circuit reliability analysis using symbolic techniques", in *IEEE Transactions on Computer-Aided Design of Integrated Circuits and Systems*, vol. 25, no. 12, pp. 2638–2649, 2006.

88. M. Zhang and N. Shanbhag, "A Soft Error rate Analysis (SERA) Methodology", in *Proceedings of ACM/IEEE International Conference on Computer Aided Design (ICCAD)*, pp. 111–118, 2004.

89. N. George, C. Elks, B. Johnson, and J. Lach, "Transient fault models and AVF estimation revisited", in *IEEE/IFIP International Conference on Dependable Systems and Networks (DSN),*pp. 477–486, 2010.

90. A. Biswas, P. Racunas, R. Cheveresan, J. Emer, S. Mukherjee, and R. Rangan, "Computing architectural vulnerability factors for address-based structures", in *Proceedings of the 32*nd *Annual International Symposium on Computer Architecture (ISCA)*, pp. 532–543, 2005.

91. N. Wang, J. Quek, T. Rafacz, and S. Patel, "Characterizing the effects of transient faults on a high-performance processor pipeline", in *IEEE International Conference on Dependable Systems and Networks (DSN)*, pp. 61–70, 2004.

92. R. Venkatasubramanian, J. Hayes, and B. Murray, "Low cost online fault detection using control flow assertions", in *Proceedings of 9th IEEE On-Line Test. Symposium (IOLTS)*, pp. 137–143, 2003.

93. P. Liden, P. Dahlgren, R. Johansson, and J. Karlsson, "On latching probability of particle induced transients in combinational networks", in *Proceedings of Fault-Tolerant Computing Symposium*, pp. 340–349, 1994.

94. J. Ziegler, H. Curtis, H. Muhlfeld, J. Montrose, and B. Chin, "IBM experiments in soft fails in computer electronics (1978–1994)", in *IBM journal of research and development*, vol. 40, no. 1, pp. 3–18, 1996.

95. L. Chen, M. Ebrahimi, and M. Tahoori, "CEP: Correlated Error Propagation for Hierarchical Soft Error Analysis", in *Journal of Electronic Testing: Theory and Applications (JETTA)*, Springer, 2013.

96. H. Ziade, R. Ayoubi, and R. Velazco, "A survey on fault injection techniques", in *International Arab Journal of Information Technology*, vol. 1, no. 2, pp. 171–186, 2004.

97. V. Chippa, D. Mohapatra, A. Raghunathan, K.Roy, and S. Chakradhar, "Scalable effort hardware design: exploiting algorithmic resilience for energy efficiency", in *Proceedings of the ACM 47th Design Automation Conference (DAC)*, pp. 555–560, 2010.

98. K. Pattabiraman, N. Nakka, Z. Kalbarczyk, and R. Iyer, "SymPLFIED: Symbolic program-level fault injection and error detection framework", in *IEEE International Conference on Dependable Systems and Networks (DSN)*, pp. 472–481, 2008.

99. R. Velazco, A. Corominas, and P. Ferreyra, "Injecting bit flip faults by means of a purely software approach: a case studied", in *IEEE International Symposium on Defect and Fault Tolerance in VLSI and Nanotechnology Systems (DFT)*, pp. 108–116, 2002.

100. J. Coppens, D. Al-Khalili, and C. Rozon, "VHDL Modelling and Analysis of Fault Secure Systems", in *Proceedings of the IEEE Conference on Design Automation and Test in Europe (DATE)*, pp. 148–152, 1998.

101. R. Shafik, P. Rosinger, and B. Al-Hashimi, "SystemC-Based Minimum Intrusive Fault Injection Technique with Improved Fault Representation", in *IEEE International On-Line Testing Symposium (IOLTS)*, pp. 99–104, 2008.

102. P. Simonen, A. Heinonen, M. Kuulusa, and J. Nurmi, "Comparison of bulk and SOI CMOS Technologies in a DSP Processor Circuit Implementation", in *Proceedings of the 13th International Conference on Microelectronics (ICM)*, pp. 107–110, 2001.

103. J. Yao, S. Okada, M. Masuda, K. Kobayashi, and Y. Nakashima, "DARA: A low-cost reliable architecture based on unhardened devices and its case study of radiation stress test", in *IEEE Transactions on Nuclear Science*, vol. 59, no. 6, pp. 2852–2858, 2012.

104. C. Weaver and T. Austin, "A fault tolerant approach to microprocessor design", in *IEEE International Conference on Dependable Systems and Networks (DSN)*, pp. 411–420, 2001.

105. G. Messenger, "Collection of Charge on Junction Nodes from Ion Tracks", in *IEEE Transactions on Nuclear Science*, vol. 29, no. 6, pp. 2024–2031, 1982.

106. P. Dodd and F. Sexton, "Critical charge concepts for CMOS SRAMs", in *IEEE Transactions on Nuclear Science*, vol. 42, no. 6, pp. 1764–1771, 1995.

107. J. Henkel, L. Bauer, H. Zhang, S. Rehman, and M. Shafique, "Multi-Layer Dependability: From Microarchitecture to Application Level", in *ACM/IEEE/EDA 51st Design Automation Conference (DAC)*, 2014.

108. C. Nguyen and G. R. Redinbo, "Fault tolerance design in JPEG 2000 image compression system", *IEEE Transactions on Dependable and Secure Computing*, vol. 2, no. 1, pp. 57–75, 2005.

109. M. A. Makhzan, A. Khajeh, A. Eltawil, and F. J. Kurdahi, "A low power JPEG2000 encoder with iterative and fault tolerant error concealment", *IEEE Transaction on Very Large Scale Integration (TVLSI)*, vol. 17, no. 6, pp. 827–837, 2009.

110. A. K. Djahromi, A. Eltawil, and F. J. Kurdahi, "Exploiting fault tolerance towards power efficient wireless multimedia applications", in *IEEE Consumer communications and networking conference*, pp. 400–404, 2007.

111. M. R. Guthaus, J. S. Ringenberg, D. Ernst, T. M. Austin, T. Mudge, and R. B. Brown, "MiBench: A free, commercially representative embedded benchmark suite", in *IEEE 4th Annual Workshop on Workload Characterization*, 2001.

112. S. Rehman, F. Kriebel, M. Shafique, and J. Henkel, "Reliability-Driven Software Transformations for Unreliable Hardware", *IEEE Transactions on Computer-Aided Design of Integrated Circuits and Systems (TCAD)*, Volume 33, Issue 11, pp. 1597–1610, 2014

113. V. Kleeberger, C. G. Dumont, C. Weis, A. Herkersdorf, D. M. Gritschneder, S. R. Nassif, U. Schlichtmann, and N. Wehn, "A Cross-Layer Technology-Based Study of How Memory Errors Impact System Resilience", *IEEE Micro*, vol. 33, no. 4, pp. 46–55, 2013.

114. S. Sinha, G. Yeric, V. Chandra, B. Cline, and Y. Cao, "Exploring sub-20 nm FinFET design with Predictive Technology Models", in *ACM/EDAC/IEEE Design Automation Conference (DAC)*, pp. 283–288, 2012.

115. F. Kriebel, S. Rehman, D. Sun, P. V. Aceituno, M. Shafique, and J. Henkel, "ACSEM: Accuracy-Configurable Fast Soft Error Masking Analysis in Combinatorial Circuits", in *IEEE/ACM 18th Design, Automation and Test in Europe Conference (DATE)*, March 2015.

116. F. Oboril, "Cross-Layer Approaches for an Aging-Aware Design of Nanoscale Microprocessors", *Ph.D. Thesis*, 2015.

117. H. Amrouch, V. M. van Santen, T. Ebi, V. Wenzel, and J. Henkel, "Towards interdependencies of aging mechanisms", in *IEEE International Conference on Computer Aided Design (ICCAD)*, pp. 478–485, 2014.

118. DFG SPP1500 Program on Dependable Embedded Systems: http://spp1500.itec.kit.edu/.

119. R. Azevedo, S. Rigo, M. Bartholomeu, G. Araujo, C. C. de Araujo, and E. Barros, "The ArchC Architecture Description Language and Tools", *International Journal of Parallel Programming*, vol. 33, no. 5, pp. 453–484, 2005.

120. C.-C. Han, K. G. Shin, and J. Wu, "A fault-tolerant scheduling algorithm for real-time periodic tasks with possible software faults", *IEEE Transactions on Computers (TC)*, vol. 52, no. 3, pp. 362–372, 2003.

121. T. Ball and J. R. Larus, "Branch Prediction for Free", *ACM SIGPLAN*, vol. 28, pp. 300–313, 1993.

122. Synopsys, "Synopsys Design Compiler User Guide" [Online]. *Available:* https://solvnet.synopsys.com/dow_retrieve/latest/dcug/dcug.html.

123. Synopsys, "Accelerate Design Innovation with Design Compiler®", Synopsys [Online]. *Available:* http://www.synopsys.com/Tools/Implementation/RTLSynthesis/Pages/default.aspx.

124. ModelSim, "ModelSim—Leading Simulation and Debugging", Mentor [Online]. *Available:* http://www.mentor.com/products/fpga/model.

125. Synopsys, "TSMC 45 nm High Speed Tapless Standard Cell Logic Library", TSMC [Online]. *Available:* http://www.synopsys.com/dw/ipdir.php?c=dwc_logic_ts45nkkslogcassst000f.

126. Flux calculator: www.seutest.com/cgi-bin/FluxCalculator.cgi.

127. http://www.archc.org; The ArchC Website.

128. http://downloads.sourceforge.net/archc/ac_lrm-v2.0.pdf; The ArchC Architecture Description Language v2.0 Reference Manual.

129. GCC: https://gcc.gnu.org/.

130. M. Shafique, L. Bauer, and J. Henkel, "Optimizing the H.264/AVC Video Encoder Application Structure for Reconfigurable and Application-Specific Platforms", *Journal of Signal Processing Systems (JSPS)*, vol. 60, no. 2, pp. 183–210, 2010.

131. *Haifa Scheduler:* http://gcc.gnu.org/, http://opensource.apple.com/ source/gcc_os/gcc_os-1660/gcc/haifa-sched.c.

132. A. Parikh, S. Kim, M. Kandemir, N. Vijaykrishnan, and M. J. Irwin, "Instruction scheduling for low-power", *Journal of VLSI Signal Processing systems*, vol 37, no. 1, pp. 129–149, 2004.

133. J. Yan and W. Zhang, "Compiler guided register reliability improvement against soft errors", in *IEEE International Conference on Embedded Software (EMSOFT)*, pp. 203–209, 2005.

134. J. Cong and K. Gururaj, "Assuring Application-Level Correctness Against Soft Errors", in *IEEE International Conference on Computer Aided Design (ICCAD)*, pp. 150–157, 2011.

135. R. Baumann, "Soft errors in advanced computer systems", in *IEEE Design & Test of Computers*, vol. 22, no. 3, pp. 258–266, 2005.

136. R. Heald, "How cosmic rays cause computer downtime", in *IEEE Reliability Society Meeting (SCV)*, pp. 15–21, 2005.

137. K. Kang, S. Gangwal, S. Park, and K. Roy, "NBTI induced performance degradation in logic and memory circuits: how effectively can we approach a reliability solution?", in *Proceedings of Asia and South Pacific Design Automation Conference*, pp. 726–731, 2008.

138. M. Shafique, M. U. K. Khan, O. Tuefek, and J. Henkel, "EnAAM: Energy-Efficient Anti-Aging for On-Chip Video Memories", in *ACM/EDAC/IEEE 52nd Design Automation Conference*, San Francisco, CA/USA, June 8–12, 2015.

139. S. Herbert, S. Garg, and D. Marculescu, "Exploiting process variability in voltage/frequency Control", *IEEE Transactions Very Large Scale Integration (VLSI) Systems*, on 20, no. 8, pp. 1392–1404, 2012.

140. T. Li, R. Ragel, and S. Parameswaran, "Reli: Hardware/software Checkpoint and Recovery scheme for embedded processors", in *IEEE Design, Automation & Test in Europe Conference & Exhibition*, pp. 875–880, 2012.

141. M. Demertzi, M. Annavaram, and M. Hall, "Analyzing the effects of compiler optimizations on application reliability", *IEEE International Symposium on Workload Characterization (IISWC)*, pp. 184–193, 2011.

142. S. Rehman, A. Toma, F. Kriebel, M. Shafique, J.-J. Chen, and J. Henkel, "Reliable Code Generation and Execution on Unreliable Hardware under Joint Functional and Timing Reliability Considerations", in: *19th IEEE Real-Time and Embedded Technology and Applications Symposium (RTAS)*, pp. 273–282, 2013.

143. J. B. Velamala, K. Sutaria, T. Sato, and Y. Cao, "Physics matters: statistical aging prediction under trapping/detrapping", in *49th IEEE/ACM Annual Design Automation Conference (DAC)*, pp. 139–144, 2012.

144. K. Kuhn, C. Kenyon, A. Kornfeld, M. Liu, A. Maheshwari, W. Shih, S. Sivakumar, G. Taylor, P. VanDerVoorn, and K. Zawadzki, "Managing Process Variation in Intel's 45 nm CMOS Technology", in *Intel Technology Journal*, vol. 12, no. 2, 2008.

145. C. Li and W. Fuchs, "Catch-compiler-assisted techniques for checkpointing", in *20th International Symposium of Fault-Tolerant Computing (FTCS-20)*, Digest of Papers, pp. 74–81, 1990.

146. J. Plank, M. Beck, G. Kingsley, and K. Li, "Libckpt: Transparent Checkpointing under Unix", in *Proceedings of Usenix Technical Conference*, pp. 213–223, 1995.

147. Y. Huang and C. Kintala, "Software implemented fault tolerance: Technologies and experience", in *Proceedings of the IEEE Fault-Tolerant Computing Symposium (FTCS)*, vol. 23, pp. 2–9, 1993.

148. L. Wang, Z. Kalbarczyk, W. Gu, and R. Iyer, "An OS-level framework for providing application-aware reliability", in *Proceedings of the 12th Pacific Rim International Symposium on Dependable Computing (PRDC)*, pp. 55–62, 2006.

149. T. Ebi, M. A. Al Faruque, and J. Henkel, "TAPE: Thermal-aware agent-based power econom multi/many-core architectures", in *IEEE International Conference on Computer Aided Design (ICCAD)*, pp. 302–309, 2009.

150. H. Khdr, T. Ebi, M. Shafique, H. Amrouch, and J. Henkel, "mDTM: Multi-Objective Dynamic Thermal Management for On-Chip Systems", in *IEEE/ACM 17th Design Automation and Test in Europe Conference (DATE)*, 2014.

151. J. Henkel, T. Ebi, H. Amrouch, and H. Khdr, "Thermal management for dependable on-chip systems", in *Asia and South Pacific Design Automation Conference (ASP-DAC)*, pp. 113–118, 2013.

152. H. Amrouch, T. Ebi, and J. Henkel, "RESI: Register-Embedded Self-Immunity for Reliability Enhancement", *IEEE Transactions on CAD of Integrated Circuits and Systems (TCAD)*, vol. 33, no. 5, pp. 677–690, 2014.

153. L. Bauer, C. Braun, M. E. Imhof, M. A. Kochte, E. Schneider, H. Zhang, J. Henkel, and H.-J. Wunderlich, "Test Strategies for Reliable Runtime Reconfigurable Architectures", in *IEEE Transactions on Computers (TC)*, vol. 62, no. 8, pp. 1494–1507, 2013.

154. H. Zhang, M. A. Kochte, M. E. Imhof, L. Bauer, H.-J. Wunderlich, and J. Henkel, "GUARD: GUAranteed Reliability in Dynamically Reconfigurable Systems", in *IEEE/ACM Design Automation Conference (DAC)*, pp. 32:1–32:6, 2014.

155. D. Gnad, M. Shafique, F. Kriebel, S. Rehman, D. Sun, and J. Henkel, "Hayat: Harnessing Dark Silicon and Variability for Aging Deceleration and Balancing", in *ACM/EDAC/IEEE 52nd Design Automation Conference (DAC)*, 2015.

156. J. Henkel, L. Bauer, J. Becker, O. Bringmann, U. Brinkschulte, S. Chakraborty, M. Engel, R. Ernst, H. Härtig, L. Hedrich, A. Herkersdorf, R. Kapitza, D. Lohmann, P. Marwedel, M. Platzner, W. Rosenstiel, U. Schlichtmann, O. Spinczyk, M. B. Tahoori, J. Teich, N. Wehn, and H. J. Wunderlich, "Design and architectures for dependable embedded systems", in *IEEE*

International Conference on Hardware/Software Codesign and System Synthesis (CODES + ISSS), pp. 69–78, 2011.

157. J. Henkel, A. Herkersdorf, L. Bauer, T. Wild, M. Hübner, R.K. Pujari, A. Grudnitsky, J. Heisswolf, A. Zaib, B. Vogel, V. Lari, and S. Kobbe, "Invasive Manycore Architectures", in *17th Asia and South Pacific Design Automation Conference (ASP-DAC)*, pp. 193–200, 2012.

158. J. Teich, J. Henkel, A. Herkersdorf, D. Schmitt-Landsiedel, W. Schröder-Preikschat, and G. Snelting, "Invasive Computing: An Overview", in *Multiprocessor System-on-Chip—Hardware Design and Tool Integration*, M. Hübner and J. Becker (Eds.), pp. 241–268, Springer, 2011.